EVOLUTION OF FOSSIL ECOSYSTEMS

Paul A. Selden BSc PhD
Reader in Palaeontology
University of Manchester, UK
President, International Society of Arachnology

John R. Nudds BSc PhD
Senior Lecturer in Palaeontology
University of Manchester, UK
Fellow of the Geological Society, London

MANSON
PUBLISHING

DEDICATION

We dedicate this book to Professor Charles H. Holland
and Professor Harry B. Whittington.

For full details of all Manson Publishing Ltd titles please write to:
Manson Publishing Ltd, 73 Corringham Road, London NW11 7DL, UK.
Tel: +44(0)20 8905 5150
Fax: +44(0)20 8201 9233
Website: www.manson-publishing.com

Commissioning editor: Jill Northcott
Project manager: Paul Bennett
Copy-editor: Kathryn Rhodes
Cover design: Patrick Daly
Book design and layout: Cathy Martin, Presspack Computing Ltd
Colour reproduction: Tenon & Polert Colour Scanning Ltd, Hong Kong
Printed by: Grafos SA, Barcelona, Spain

CONTENTS

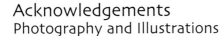

Acknowledgements
Photography and Illustrations

American Museum of Natural History Library: 158, 161.

Cristoph Bartels, Deutsches Bergbau-Museum, Bochum: 49, 50, 63, 65, 66, 70.

Fred Broadhurst, University of Manchester: 97, 98, 100, 101, 102, 103, 104, 105, 106, 107, 108, 109, 110, 111, 112, 113, 114, 115, 116.

Simon Conway Morris, University of Cambridge: 25, 27, 28, 36, 38.

Richard Fortey, Natural History Museum, London: 44.

Sarah Gabbott, University of Leicester: 43, 45, 46, 47.

Jean-Claude Gall, Université Louis Pasteur de Strasbourg: 121, 122, 123, 124, 125, 127, 128, 129, 130, 131, 132.

David Green, University of Manchester: 30, 60, 185, 186, 192, 203, 206, 209.

Richard Hartley, University of Manchester: 1, 4, 5, 12, 18, 19, 22, 39, 41, 48, 53, 77, 78, 83, 84, 85, 86, 87, 88, 91, 92, 93, 94, 95, 96, 99, 117, 118, 120, 133, 136, 153, 155, 174, 179, 193, 194, 219, 232, 233, 249, 252.

Rolf Hauff, Urwelt-Museum Hauff: 137, 139, 140, 142, 144, 145, 147, 150, 151, 152.

David Martill, University of Portsmouth: 201, 213, 214, 215, 216, 218.

Barbara Mohr, Humboldt Museum Berlin: 208.

Natural History Museum of Los Angeles County: 254, 257, 259, 266.

John Nudds, University of Manchester: 2, 3, 6, 8, 11, 15, 20, 21, 51, 52, 55, 57, 58, 59, 61, 68, 134, 135, 154, 156, 157, 165, 168, 169, 171, 175, 176, 195, 196, 197, 198, 199, 200, 202, 204, 205, 250, 251, 253, 262, 264.

Burkhard Pohl, Wyoming Dinosaur Center: 167.

Glenn Rockers, PaleoSearch, Kansas: 188.

Graham Rosewarne, Avening, Gloucestershire: 23, 24, 26, 29, 31, 32, 33, 35, 37, 54, 56, 62, 64, 67, 69, 138, 141, 143, 146, 148, 149, 160, 163, 166, 170, 173, 180, 182, 183, 184, 187, 191, 210, 212, 217, 255, 256, 258, 260, 261, 263, 265.

Sauriermuseum Aathal, Switzerland: 159, 162, 172.

Paul Selden, University of Manchester: 7, 9, 10, 13, 14, 16, 17, 40, 42, 71, 72, 73, 74, 75, 76, 79, 80, 81, 82, 89, 90, 119, 126, 207, 230, 231.

Forschungsinstitut und Naturmuseum Senckenberg, Messel Research Department: 220, 221, 222, 223, 224, 225, 226, 227, 228, 229.

Geoff Thompson, University of Manchester: 177, 178, 181, 189, 190, 211.

The University of Wyoming Geological Museum: 164.

Wolfgang Weitschat, University of Hamburg, Germany: 234, 235, 236, 237, 238, 239, 240, 241, 242, 243, 244, 245, 246, 247, 248

H.B.Whittington, University of Cambridge: 34.

Access to sites and other help

Artur Andrade, Brent Breithaupt, Paulo Brito, Des Collins, John Dalingwater, Mike Flynn, Jim Gehling, Zhouping Guo, Rolf Hauff, Andre Herzog, Ken Higgs, Mary Howie, Neal Larson, Bob Loveridge, Terry Manning, David Martill, Cathy McNassor, Urs Möckli, Robert Morris, Sam Morris, Robert Nudds, Jenny Palling, Burkhard Pohl, Helen Read, Glenn Rockers, Chris Shaw, Bill Shear, Roger Smith, Wouter Südkamp, Hannes Theron, Rene Vandervelde, Jane Washington-Evans.

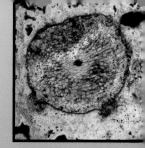

PREFACE

Most major advances in understanding the history of life on Earth in recent years have been through the study of exceptionally well-preserved biotas (Fossil-Lagerstätten). Indeed, particular Fossil-Lagerstätten, such as the Burgess Shale of British Columbia and the Solnhofen Limestone of Bavaria, have gained exceptional fame through popular science writings. Study of a selection of such sites scattered throughout the geological record – windows on the history of life on Earth – can provide a fairly complete picture of the evolution of ecosystems through time.

This book arose from the realization that there was an obvious gap in the current range of palaeontology texts which would be filled by a book which brings together succinct summaries of most of the better-known Fossil-Lagerstätten, primarily for a student and interested amateur readership. The authors teach an undergraduate course at third-year level which is based on case studies of a number of Fossil-Lagerstätten, and have also collaborated on the design of a new fossil gallery at Manchester University Museum based around this theme.

Following an introduction to Fossil-Lagerstätten and their distribution through geological time, each succeeding chapter deals with a single fossil locality. Each has the same format: after a brief introduction placing the Lagerstätte in an evolutionary context, there then follows a history of study of the locality; the background sedimentology, stratigraphy and palaeoenvironment; a description of the biota; discussion of the palaeoecology; and a comparison with other Lagerstätten of a similar age and/or environment, and suggestions for further reading. An Appendix gives information about museums to visit which display fossils of each locality, and suggestions for visiting the sites.

Abbreviations

Specimen repositories are abbreviated in the figure captions as follows:

AMNH	American Museum of Natural History, New York
BKM	Bad Kreuznach Museum, Germany
BM	Bundenbach Museum, Germany
BSPGM	Bayerische Staatssammlung für Paläontologie und Historische Geologie, München, Germany
CFM	Field Museum, Chicago
DBMB	Deutsches Bergbau-Museum, Bochum, Germany
GCPM	George C. Page Museum, Los Angeles
GGUS	Grauvogel–Gall Collection, Université Louis Pasteur, Strasbourg, France
GMC	Geological Museum, Copenhagen, Denmark
GPMH	Geologische-Paläontologisches Museum, Universität Hamburg, Germany
GSSA	Geological Survey of South Africa
HMB	Humboldt Museum Berlin, Germany
MM	Manchester University Museum, UK
MU	Manchester University Earth Sciences Department, UK
NHM	Natural History Museum, London, UK
PC	Private Collection
SCM	Santana do Cariri Museum, Brazil
SI	Smithsonian Institute, Washington DC
SM	Simmern Museum, Germany
SMA	Sauriermuseum Aathal, Switzerland
SMFM	Senckenberg Museum, Frankfurt-am-Main, Germany
SMNS	Staatliches Museum für Naturkunde, Stuttgart, Germany
UMH	Urwelt-Museum Hauff, Holzmaden, Germany
UWGM	University of Wyoming Geological Museum, Laramie, Wyoming
WAM	Western Australian Museum, Perth, Western Australia
WDC	Wyoming Dinosaur Center, Thermopolis, Wyoming

INTRODUCTION

The fossil record is spectacularly incomplete. Only a tiny proportion of plants and animals alive at any one time gets preserved as fossils, so that the palaeontologist attempting to reconstruct ancient ecosystems is, in effect, trying to complete a jigsaw puzzle without the picture on the box lid and for which the majority of pieces are missing. Under normal preservational conditions probably only around 15% of organisms are preserved. Moreover, the fossil record is biased in favour of those animals and plants with hard, mineralized shells, skeletons or cuticle, and towards those living in marine environments. Thus, the *preservational potential* of a particular organism depends on two main factors: its constitution (better if it contains hard parts), and its habitat (better if it lives in an environment where sedimentary deposition occurs).

Occasionally, however, the fossil record presents us with surprises. Very rarely, exceptional circumstances of one sort or another allow unusual preservation of soft parts of organisms, or in environments where fossilization rarely happens. Rock strata within the geological record which contain a much more completely preserved record than is normally the case are windows on the history of life on Earth. They have been termed Fossil-Lagerstätten (Seilacher *et al.*, 1985), a name derived from German mining tradition to denote a particularly rich seam.

There are two main types of Fossil-Lagerstätten. *Concentration Lagerstätten* (Konzentrat-Lagerstätten), as the name suggests, are simply deposits in which vast numbers of fossils are preserved, such as coquinas (shell accumulations), bone beds, cave deposits and natural animal traps. The quality of individual preservation may not be exceptional but the sheer numbers are informative. *Conservation Lagerstätten* (Konservat-Lagerstätten), on the other hand, preserve quality rather than quantity, and this term is restricted to those rare instances where peculiar preservational conditions have allowed even the soft tissue of animals and plants to be preserved, often in incredible detail. Most of the Lagerstätten described in this book are examples of Conservation Lagerstätten.

There are various types of Conservation Lagerstätten including *conservation traps* such as entombment in amber, deep-freezing in permafrost, pickling in oil swamps and mummification by desiccation. On a larger scale are *obrution deposits*, where episodic smothering ensures rapid burial of mainly benthic (sea-floor) communities, and *stagnation deposits*, where anoxic (low oxygen) conditions in stagnant or hypersaline (high salinity) bottom waters ensure reduced microbial decay, in predominantly pelagic (open-sea) communities. In fact, most Conservation Lagerstätten combine obrution and stagnation in the preservation of soft tissue.

Taphonomy is the name given to the process of preservation of a plant or animal as a fossil. It actually consists of two main processes. The first is *biostratinomy*, which covers the course of events from death to burial in sediment (or entombment in amber, cave deposits, and so on). The time taken by this process can vary from a few minutes (e.g. insects trapped in amber, or mammals in tar) to many years for an accumulation of bones or shells. Ideally, for exceptional preservation of soft tissues, the time between death and isolation from oxygen and decaying organisms should be short. Following burial, the second process of *diagenesis* begins, which is the conversion of soft sediment or other deposits to rock. Further destruction of organic molecules can occur during diagenesis; for example, the action of heat can turn organic molecules into oil and gas, and crushing in coarse sand can fragment plant and animal cuticles.

Soft tissue preservation has three important implications. First, the study of soft-part morphology alongside the morphology of the shell or skeleton allows better comparison with extant forms and provides additional phylogenetic information. Second, it enables the preservation of animals and plants which are entirely soft-bodied and which would normally stand no chance of fossilization. For example, it has been estimated that 85% of the Burgess Shale genera (Chapter 2) were entirely soft-bodied and are therefore absent from Cambrian biotas preserved under normal taphonomic conditions. The third implication follows –

such Conservation Lagerstätten therefore preserve for the palaeontologist a complete (or much more nearly complete) ecosystem. Comparison of such horizons in a chronological framework gives us an insight into the evolution of ecosystems over geological time.

The Lagerstätten described in this book are arranged in chronological order, from the late Pre-cambrian Ediacara biota to the Pleistocene Rancho La Brea (**Table 1**), so it is possible to follow the development of the Earth's ecosystems through a series of snapshots of life at a number of points in time (see also Bottjer *et al.*, 2002). Although they do not give a complete picture of the evolving biosphere, Lagerstätten are important because they preserve far more of the biota than occurs under normal preservational conditions, so the palaeontologist can see more completely the ecological interactions of the organisms in that particular habitat. In the late Precambrian Ediacara biota, for example, we may be looking at a different grade of organismal organization and life mode than we see in later, Phanerozoic time. This biota existed before hard parts of animals evolved and predation became widespread. By the middle of the Cambrian Period, almost all animal phyla had developed, and it is possible to reconstruct the ecological dynamics of the sea floor, complete with predators, scavengers, filter feeders and deposit feeders. A similarly diverse assemblage is seen in the Devonian Hunsrück Slate. Ediacara, the Burgess Shale and the Hunsrück Slate all preserve mainly benthos (sea-floor dwellers). In the Burgess Shale these can be divided into infauna (animals living within the sediment) and epifauna (animals living on the sediment surface). Some of the biota in the Burgess Shale and Hunsrück Slate is part of the nekton (swimmers); and in the Ordovician Soom Shale the nekton is dominant because there were only rare occasions when the sea floor was conducive to benthic life. To complete the picture of marine lifestyles, organisms which float are called plankton.

A major evolutionary advance which occurred in the mid-Palaeozoic was the colonization of land by plants and animals, and the Devonian Rhynie Chert was one of the first-known – and is still the best-known – biotas, preserving some of the earliest land plants and animals. By the late Carboniferous the land in tropical regions had become well colonized by forest, with its accompaniment of insects and their predators. The Mazon Creek biota preserves a forest ecosystem mingled with non-marine aquatic organisms in a deltaic setting, so common in this period of major coal formation. The end of the Permian Period saw the demise of some 80% of living things in the greatest mass extinction of all time,

yet when we examine the biota of the Triassic Grès à Voltzia delta, we find many similarities to that of the Mazon Creek. Three Lagerstätten are represented from the Jurassic Period, two marine and one terrestrial. The Holzmaden *Posidonia* Shales represent a snapshot of marine pelagic life in the Jurassic Period, in which large marine vertebrates such as plesiosaurs, ichthyosaurs and crocodiles are found together with their prey: cephalopods and fish. In contrast, the Solnhofen Plattenkalk (a German word meaning 'platy limestones') preserves marine plankton, nekton and benthos (e.g. ammonites, horseshoe crabs, crustaceans) as well as rare flying animals (e.g. *Archaeopteryx* – the first bird), all swept together into a lagoon by storms. On land in the Jurassic Period, dinosaurs dominated the scene, and the Morrison Formation of the western USA is the best-known Lagerstätte preserving these giants.

During the Cretaceous Period, a region of what is now north-eastern Brazil witnessed not one but two Fossil-Lagerstätten: the Santana and Crato Formations. The former is best known for its fish and pterosaurs in nodules, the latter for its insects and plants in a Plattenkalk. The dinosaurs died out together with ammonites, marine reptiles and some other plant and animal groups at the end of the Cretaceous Period. In the following Cenozoic Era, mammals became the dominant vertebrate group, and some of the finest fossils of these, together with a forest-dwelling flora and fauna, occur at Grube Messel in Germany. Land animals and plants are generally much rarer as fossils than those which lived in places where sediments were being laid down, such as lakes and the sea, so Lagerstätten which preserve terrestrial biotas are especially prized. Grube Messel is one such, and the amazing fauna (especially insects) of Baltic amber is another. Amber (fossilized tree resin) acts as a sticky trap for insects and their predators and, in a similar manner, the tar pits of Rancho La Brea attracted mammals and birds in search of a drink, trapping them and their predators and scavengers in sticky tar. Rancho La Brea thus preserves a snapshot of land life in southern California over the last 40,000 years.

Further Reading

Bottjer, D. J., Etter, W., Hagadorn, J. W. and Tang, C. M. (eds.). 2002. *Exceptional fossil preservation*. Columbia University Press, New York, xiv + 403 pp.

Seilacher, A., Reif, W-E. and Westphal, F. 1985. Sedimentological, ecological and temporal patterns of fossil Lagerstätten. *Philosophical Transactions of the Royal Society of London*, Series B **311**, 5–23.

Million years before present	Era	Period			Lagerstätten
2.5	CENOZOIC	Quaternary		Holocene	Rancho La Brea
				Pleistocene	
23.5		Tertiary	Neogene	Pliocene	
				Miocene	Baltic Amber
			Palaeogene	Oligocene	Grube Messel
				Eocene	
				Palaeocene	
65					
146	MESOZOIC	Cretaceous			Santana & Crato
		Jurassic			Solnhofen Limestone
					Morrison Formation
205					Holzmaden Shales
		Triassic			Grès à Voltzia
251					
290	PALAEOZOIC	Permian			
		Upper Carboniferous (Pennsylvanian)			Mazon Creek
320		Lower Carboniferous (Mississippian)			
353					
		Devonian			Hunsrück Slate
409					Rhynie Chert
		Silurian			
439		Ordovician			Soom Shale
510					
		Cambrian			Burgess Shale
540					
	PRECAMBRIAN				Ediacara
4600					

Table 1 The Geological Column showing stratigraphic positions of the Fossil–Lagerstätten described in this book.

Chapter One

EDIACARA

Background: first life on Earth

Life on Earth arose some 3,500 million years ago. There is some debate concerning what actually constitutes 'life' and, indeed, whether life actually arose on this planet or originated extraterrestrially in a simple form and further evolved here. Nevertheless, the earliest fossil evidence of single-celled prokaryotes akin to modern cyanobacteria (blue-green algae) comes from cherts in Western Australia. For some 2,500 million years after its origin, life evolved slowly. Eukaryotes (cells which contain a nucleus and organelles) evolved from prokaryotes, but it was not until about 1,000 million years ago that multicellular forms developed. These first multicellular organisms form the subject of this chapter. Whether they are plants (metaphytes), animals (metazoans) or neither is unknown, but they were typically flat creatures, with a high surface area/body mass ratio. The development of multicellularity was a major step in the evolution of life: it enabled organisms to grow in size, to develop organ systems through tissue differentiation, and led to the plants and animals we are familiar with today.

Until the middle of the last century it was thought that rocks older than Cambrian in age, collectively called Precambrian, were devoid of fossils of multicellular creatures. The base of the Cambrian is clearly marked by the sudden appearance of shelly fossils – brachiopods, trilobites and sponges, for example. The discovery of soft-bodied organisms similar in appearance to jellyfish and worms in rocks of late Precambrian age was therefore a major surprise, and led to a complete reappraisal not only of the fossil record of multicellular organisms but also of the evolution of life and its relationship to the Earth's physical systems (atmosphere, oceans). The question changed from 'why did multicellular life suddenly appear at the base of the Cambrian?' to 'why did multicellular organisms suddenly develop hard parts at the start of the Cambrian?' (see Chapter 2).

History of discovery of the Ediacara biota

In 1946 Reginald C. Sprigg, a government geologist, was exploring an area of the Flinders Ranges some 300 km (190 miles) north of Adelaide, Australia, known as the Ediacara Hills (1). In the Ediacara Hills he found fossilized imprints of what were apparently soft-bodied organisms, preserved mostly on the undersides of slabs of quartzite and sandstone (2, 3). Most were round, disc-shaped forms that Sprigg called 'medusoids' from their seeming similarity to jellyfish (Sprigg, 1947, 1949). Others resembled worms and arthropods, and some could not be classified.

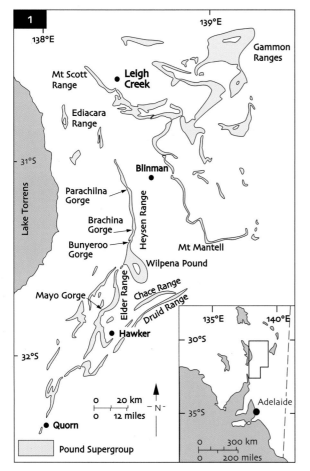

1 Map showing distribution of the Pound Supergroup in South Australia (after Gehling, 1988).

2 Greenwood Cliff in the Ediacara Hills, the site of Sprigg's discovery of fossils in the Rawnsley Quartzite in 1946.

3 Overturned slabs of Rawnsley Quartzite, Ediacara Hills, with fossils preserved on the rippled undersurfaces.

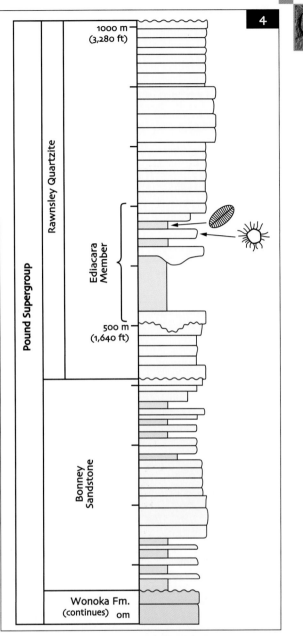

4 Stratigraphy of the late Proterozoic Pound Supergroup of South Australia, showing the position of the Ediacara biota in the Ediacara Member (after Bottjer, 2002).

Initially, Sprigg thought that these rocks were Cambrian in age because they contained fossils, but later work established that their age was, in fact, late Precambrian. While scattered reports of soft-bodied organisms had appeared in the scientific literature as far back as the mid-nineteenth century, this was the first diverse assemblage of well-preserved Precambrian fossils to be discovered and was studied in detail by Martin Glaessner and Mary Wade of the University of Adelaide (Glaessner and Wade, 1966). Not long after Sprigg's discovery, assemblages of soft-bodied organisms were discovered in Leicestershire, UK (Ford, 1958) and Namibia, and Ediacaran-type fossils are now known from the White Sea area of Russia, Newfoundland and Northwest Territories (Canada), North Carolina (USA), Ukraine, China, and many other places. They all occur in the period from approximately 670 million years ago to the earliest Cambrian (about 540 million years ago). This period of time has been named the Ediacaran (or Vendian) period, and is characterized by the presence of these fossils.

Stratigraphic setting and taphonomy of the Ediacara biota

The first fossils found by Sprigg came from the Ediacara Hills area, but rock sequences containing Precambrian fossils also occur in gorges through the Heysen Range to the south (e.g. Parachilna Gorge, Brachina Gorge, Bunyeroo Gorge, Mayo Gorge, 1), and at the eastern end of the Chace Range. The fossils are confined to a stratigraphic range of no more than 110 m (360 ft) in the Ediacara Member of the Rawnsley Quartzite, which lies 500 m (1,640 ft) below the earliest Cambrian in this area. The Rawnsley Quartzite is part of the Pound Supergroup (4), named after Wilpena Pound, a dramatic eroded syncline whose circle of quartzite cliffs faces outwards in a natural fortification.

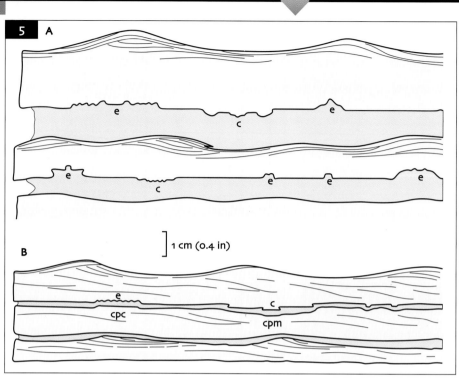

5 Preservational styles in the Ediacara Member; **A** with thick and **B** with thin clay interlayers. **c** cast on base of sandstone; **e** external mould on base of sandstone, **cpc** counterpart cast on top of sandstone, **cpm** counterpart mould on top of sandstone (after Gehling, 1988).

1 cm (0.4 in)

6 Microbial mat preserved on the surface of ripples, Rawnsley Quartzite, Ediacara Hills.

The Ediacara Member consists of a series of silt-stones and sandstones which represent pelagic to inter-tidal conditions. The implication is that there was a continental edge delivering sediment into deep water, at times by means of turbidite flows and occasionally as a delta which shallowed the water to sub- and inter-tidal levels. Some storm horizons can be seen. It is at these shallow levels, around the storm wave base, that the fossils occur. Aiding preservation are the thin films of clay which represent gentle deposition from suspension and which occur between the sandstone layers, the latter representing more energetic flow and sediment

deposition, possibly during storm events. The clay acted to some extent as a glue, cohering the sands beneath and the fossils lying on the sea bed, and it moulds fine detail of the fossils, enabling interpretation of their morphology. Some of the rippled surfaces also show evidence of a microbial mat on the sea bed (6), which would have also aided preservation by enclosing carcasses in a closed environment.

Being soft-bodied, the fossils are generally preserved squashed. The clay layers compact considerably during diagenesis, so relief is provided by the sandstones. Figure 5 (from Gehling, 1988) shows the effect of different thicknesses of clay on the preservation of the fossils. In some cases, an external mould of the upper surface of the fossil is preserved on the base of the sandstone as an impression (e in 5). Sometimes, the fossil collapses or decays so that sand fills the space previously occupied by the organism and produces a cast, visible as a positive relief on the base of the sandstone (c in 5). If the clay layers are thin (5B) then the cast can project deeper into the soft sand beneath, forming a counterpart mould (cpm in 5B); conversely, a counterpart cast (cpc in 5B) can also form. In this way, dorsal and ventral structures may be superimposed upon one another. Some of the organisms had thin outer walls but more resistant internal organs which are moulded preferentially, e.g. the gonads of possible medusoids. A more recent interpretation of the preservation of the biota (Gehling, 1999) recognized the contribution of microbial mats.

Because the fossils are only moulds and casts, no organic matter remains: they are best viewed in low-angle light, either evening light in the field or light from a lamp at low angle in the laboratory. Silicone rubber casts of fossils preserved as moulds, and vice versa, can provide better views of some material.

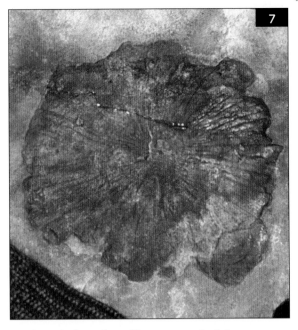

7 Giant *Cyclomedusa*. About 30 cm (12 in) across.

8 Smaller *Cyclomedusa*, from the locality shown in **3**. Coin 24 mm (0.94 in) diameter.

Description of the Ediacara biota

Ediacaria is a concentric discoidal form first described by Sprigg in the 1940s. It has now been found in other Ediacaran localities, e.g. north-west Canada. It could be a jellyfish, a benthic form, or even the holdfast of another organism.

Cyclomedusa (**7**, **8**) is a primarily radiate form but with some concentric lines near the centre. It grew to nearly 1 m (3 ft) across. It is probably the commonest and most widespread of Ediacaran fossils. Early authors suggested that *Cyclomedusa* was a large, floating jellyfish; Seilacher (1989) envisaged it as benthic; it could be interpreted as a low, cone-like form of sea anemone (Gehling, 1991); and another hypothesis is that it is simply the holdfast of a colonial octocoral. The evidence for the sea anemone form comes from the concentric lines near the centre, which could be artefacts of compression, and the occurrence of a number of specimens closely adpressed which would be unlikely if they were free-floating creatures. However, its abundance supports the theory that it was a holdfast.

Pseudorhizostomites (**9**) is a radial form with an indefinite edge. It could represent a medusoid, or possibly the impression of the base of a larger organism.

Tribrachidium is a small (c. 20 mm [0.8 in] in diameter) disc-like form with a three-fold radial symmetry consisting of three lobes in the central region, each with a raised leading edge which turns at the outer region of the central area to run along the edge of the area. There

9 Cast of *Pseudorhizostomites* (MM). About 50 mm (2 in) across.

10 Cast of *Tribrachidium* (MM). About 20 mm (0.8 in) across.

11 *Inaria*, from the locality shown in **3**. Coin 23 mm (0.9 in) diameter.

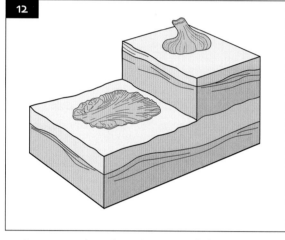

12 Reconstruction of *Inaria* (top) and sketch of fossil as it appears in the rock (below) (after Gehling, 1988).

13 Cast of *Charnia* (MM). Length about 150 mm (6 in).

is an outer zone composed of three flat areas which appear to emanate from the central lobes; each area bears radiating ridges (**10**). *Tribrachidium* does not fit easily in any extant phylum.

Mawsonites is another radial form, typified by concentric rows of lobate shapes getting larger from the centre outwards. It was interpreted as a medusoid by Glaessner and Wade (1966), but as a complex trace fossil by Seilacher (1989).

Arkarua was described by Jim Gehling from the Chace Range. Its five-fold symmetry immediately suggests it may be an echinoderm – the earliest known – which has yet to develop the calcareous plates typical of most modern members of the phylum.

Inaria would fit with radial forms, except that reconstruction of the organism shows it to have been shaped rather like a garlic bulb in life (**11**, **12**). Its describer, Jim Gehling (1988), suggested that it might be a type of sea anemone.

Charnia (**13**) is one of two fossils described by Trevor Ford in 1958 from Charnwood Forest, Leicestershire; the other is *Charniodiscus*. *Charnia* is a leaf-shaped structure with a central, apparently stiff shaft and lateral segmented areas, while *Charniodiscus* resembles *Ediacaria* in being a simple disc with concentric lines. Later, a specimen was found in which a *Charnia* was attached to a *Charniodiscus*, and it is now certain that *Charniodiscus* is the root or holdfast of the leaf-like *Charnia*. *Charnia* appears to be a sea pen, a group of Cnidaria which also occur in the Burgess Shale (Chapter 2: *Thaumaptilon*) and at the present day.

Rangea is similar to *Charnia* but shows subsegmentation of the individual segments on each side.

Pteridinium is another sea pen-like organism without subsegmentation.

Ernietta was sock-shaped in life; its segments were directed upwards and its base appears to have been filled with sediment as an anchor.

Kimberella was originally considered to be a medusoid with four-fold symmetry, but later work suggested that it was benthic and slug-like, possibly an early mollusc.

Small *Dickinsonia* (**14**) look almost like the radial organisms, but as they grew (up to 1 m [3 ft] in length) they acquired an elongate body and bilateral symmetry. *Dickinsonia* is segmented on each side of a midline, but there is no head or tail end. Their appearance is reminiscent of a flatworm, except that the segments do not match up on either side (**15**).

Parvancorina (**16**) is a small creature with bilateral symmetry and a definite 'head' end.

Praecambridium is another small, bilateral organism with a 'head' end and a segmented 'body'. Considered by some to be arthropod-like, Birket-Smith (1981b) suggested it might be a juvenile *Spriggina*.

Like *Parvancorina* and *Praecambridium*, *Vendia* is another small organism with bilateral symmetry, a definite head and body segments.

Spriggina (**17**) is a small (c. 50 mm [2 in]) organism with a horseshoe-shaped 'head' followed by an elongate, leaf-like body composed of two rows of short segments either side of a medial line. At first, *Spriggina* was thought to resemble a polychaete worm such as *Nereis*, but a close look at the segmentation reveals that the segments do not match across the mid-line, just as in *Dickinsonia*. Seilacher (1989) turned the interpretation upside-down, suggesting that *Spriggina* could be another type of sea pen, and that the 'head' was actually a holdfast.

One additional fossil type present in the Ediacaran biotas is the trace fossil: evidence of crawling and shallow ploughing through the sediment by animals is present in

15 Larger *Dickinsonia*, from the locality shown in Figure **3**. Coin 24 mm (0.94 in) diameter.

14 A small *Dickinsonia* (MM). Length about 50 mm (2 in).

16 Cast of *Parvancorina* (MM). Length about 30 mm (1.2 in).

17 Cast of *Spriggina* (MM). Length about 50 mm (2 in).

many of the Ediacaran localities. For example, a spiralling trail found at Ediacara suggests grazing on sediment rather than simply getting from A to B. Though reflecting fewer body plans than seen at the present day (there are no burrowers), the traces provide evidence for true Metazoa.

First attempts to classify the Ediacaran biota, by Sprigg (1947, 1949), Glaessner (1961) and Glaessner and Wade (1966), suggested a variety of extant phyla, including Cnidaria and Annelida. Fifteen species of Ediacaran organisms were described as jellyfish (medusoids), e.g. *Mawsonites*, *Cyclomedusa*, *Kimberella*, and *Eoporpita* (the only one which shows tentacles). *Charnia*, *Pteridinium* and *Rangea* were referred to the Pennatulacea or sea pens, or soft corals, another class of Cnidaria. *Dickinsonia* and *Spriggina* were allied to the Annelida. However, other organisms such as *Tribrachidium* could not be placed in any modern phylum. Its three-fold symmetry is not seen in the ground-plan of any living group of animals.

Seilacher (1989) removed the Ediacaran organisms from the Metazoa altogether, preferring instead to regard them as a separate kingdom – Vendozoa – based on different functional design from plants and animals. He looked at the constructional morphology of the organisms and suggested they may have had internal hydrostatic skeletons, their bodies gaining rigidity through internal pressure, as in a car tyre. Compartmentalization of the body made sense to prevent total loss of pressure (and hence death) when only one or two segments were punctured. They were quilted organisms, similar to inflatable mattresses. The high surface area to mass ratio suggested they respired through the skin; they may have been photosynthetic, as suggested by McMenamin (1998), used photosynthetic symbionts like modern corals, perhaps were chemosymbiotic (with chemosynthetic symbionts so they could survive in deep-water, reducing environments), or may have ingested materials through the body wall. Seilacher suggested that such organisms could survive without bony skeletons or shells because there was little predation.

In the 1980s Fedonkin devised a classification scheme for the Ediacaran fossils based on their symmetry and thus independent of any biological interpretation. He divided the organisms into two major groups: Radiata (disc-shaped, no bilateral symmetry) and Bilateria (bilateral symmetry apparent). Radiata can be further subdivided into: Cyclozoa, with a concentric (rather than radial) pattern (e.g. *Ediacaria*); Inordozoa, with a radial pattern of indeterminate radius (e.g. *Cyclomedusa*); and Trilobozoa with a three-rayed symmetry (e.g. *Tribrachidium*). Other radiate forms were considered to belong to the cnidarian classes Conulata or Scyphozoa. Bilateria may show no particular head or tail (i.e. bidirectional, e.g. *Dickinsonia*), or have a distinct head or rooting structure (i.e. unidirectional, e.g. *Spriggina*). Intermediates exist, for example, between the concentric and indeterminate radial forms: *Cyclomedusa* shows a concentric pattern, especially near the centre of the disc. It is possible to envisage evolutionary trends among the organisms. Inordozoa (e.g. *Cyclomedusa*) could have given rise to more organized, determinate forms (e.g.

Tribrachidium), or become elongate (e.g. *Dickinsonia*) and then developed a head and tail when directed locomotion evolved (e.g. *Spriggina*). One problem with this scheme, as pointed out by Gehling, is that it only works if the organisms are flattened, yet many reconstructions show the creatures as quite three-dimensional. *Inaria*, for example, was probably bulb-shaped, and *Pteridinium* appears to have had three 'wings' rather than the two shown by the similar *Charnia*.

In a strange twist to an already bizarre story, Greg Retallack, of the University of Oregon, looked at the indeterminate growth pattern of many of the Ediacaran organisms, which suggested that they had no upper limits nor definite bounds to their size and shape (Retallack, 1994). For example, *Dickinsonia* appeared to have no 'adult' size and shape. He also studied taphonomic aspects of the Ediacaran biota, particularly their compression. The conclusion? That Ediacaran fossils were lichens! Lichens are composite organisms formed by a symbiotic relationship between green algae (which provide photosynthesis) and fungi (which provide bulk). Few palaeontologists have accepted the arguments presented by Retallack. While some Ediacaran organisms (e.g. *Spriggina*) show a distinct holdfast (or head), and their growth patterns may be less determinate than worms, for example, they are not random; and the sheer size and bulk of many forms is quite unlike that of modern lichens. Moreover, modern lichens are terrestrial, not marine. The general consensus has always been that the organisms present were metazoan-grade animals; it is indeed possible that some of the frond-like forms actually represent macro-plants.

A study by Dewell *et al.* (2001) suggested that most Edicaran animals were colonial forms (pennatulaceans such as *Charnia* are colonial), and that they represented a grade of organization more developed than simple sponges, in which few cell types are present, but less evolved than higher Eumetazoa, in which tissues and organs have developed. Thus, the Ediacarans are probably not a separate kingdom but part of the Animalia during an early phase of the development of the metazoan body plans that are familiar today.

Palaeoecology of the Ediacara biota

Early attempts to reconstruct the Ediacaran scene (e.g. Glaessner, 1961) showed a preponderance of medusoids and sea-pens in a shallow-water habitat. Later evidence, discussed above, suggests deeper water and different life styles. Indeed, Gehling (1991), in a thought-provoking essay on the Ediacaran biota, dispelled a number of earlier myths which had grown up about it. First, Ediacaran organisms were originally though it to have been large in comparison to Cambrian faunas; in fact, while some large *Dickinsonia* do occur, most specimens are small. The preoccupation with considering the Ediacaran fossils as essentially two-dimensional does not take into account preservational factors; after restoration, many organisms are conical, hemispherical or tubular. Early reconstructions showed medusoids dominating the scene; most organisms were probably benthic sessile or vagile forms. One generalization was that the assemblage was allochthonous (drifted into the site of deposition); while this may be true for some localities (e.g. Namibia), in most cases the fossils appear to represent autochthonous assemblages with just a few pelagic forms. The trace fossils described (e.g. by Crimes) were originally thought to have been formed by taxa unrepresented in the body fossil record; Gehling has suggested that a number of traces were formed by animals present in the biota. Studies on the biomechanics of Ediacaran organisms (Schopf and Baumiller, 1998) indicated that the flat animals would have been prone to dislodgement from the substrate in the strengths of currents suggested by the sedimentological evidence. They surmised that the animals might have been moved from a quieter environment to their final resting place (in which case why are they so well preserved?), were more dense or more adhesive to the substrate than other researchers had reckoned, and/or lived buried to some degree within the substrate.

Comparison of Ediacara with other late Precambrian biotas

The map in Figure **18** shows how widespread Ediacaran biotas are. The map also shows the distribution of Cambrian Burgess Shale-type faunas (Chapter 2). While many of the organisms present in the Ediacaran biotas are similar, there are some important differences between the localities, in taphonomy for example. At Ediacara itself, as discussed, the biota are preserved in shale partings in sandstone beds, in an apparently quiet-water environment. At Mistaken Point, Newfoundland, the organisms occur at the bases of volcanic ash layers, while the Namibian fossils show evidence of moving sand and mud during their deposition. Ediacaran biotas have also been found in Russia and the Ukraine (White Sea, Podolia, southern Ukraine, Ural Mountains, Siberia); North-west Canada; Central Australia; Finnmark, Norway; Charnwood Forest, Leicestershire, UK; South and North China; South-western United States and northern Mexico; and North Carolina. Restricted or doubtful Ediacaran-type biotas have also been reported from India, Iran, Ireland, Morocco, Sardinia and Pembrokeshire.

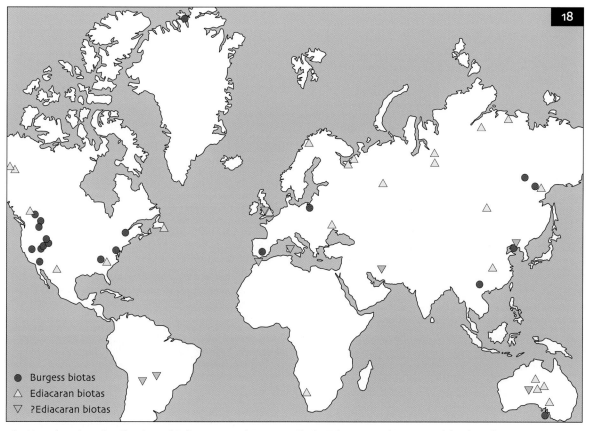

18 Map showing distribution of Ediacaran and Burgess Shale (Chapter 2) biotas worldwide (after Conway Morris, 1990).

What happened to the Ediacaran biota after the end of the Precambrian? Some apparently continued to the present day. For example, Pennatulacea occur in the Burgess Shale (Chapter 2: *Thaumaptilon*) and persist to the present day; Scyphozoa (jellyfish) are an important component of the modern marine plankton. However, there is little doubt that many forms did not survive to the Phanerozoic. At the very beginning of the Cambrian, in the Tommotian, there is a distinctive fauna of 'small shelly fossils' which just pre-date the first Cambrian skeletal body fossils. Some of these forms clearly belong to extant phyla, e.g. Mollusca; others (e.g. hyolithids) did not persist beyond the Cambrian. More interesting are the tiny spines and pastille-shaped sclerites, commonly phosphatic, which are likely to have studded the cuticle of essentially soft-bodied animals like onychophorans or halkieriids (see Chapter 2). The relationship between the Ediacara biota and the small shelly fossils is not entirely resolved.

Regardless of what the Ediacaran animals/plants/Vendozoa were, or how they lived, the shallow marine scene on Earth in late Precambrian times was utterly unlike any habitat on the planet today. It is only to be expected that, as more fossils of Ediacaran organisms are described from different parts of the world, and new ideas as to their affinities and life-styles arise, the debate about life in the Precambrian will continue.

Further Reading

Birket-Smith, S. J. R. 1981a. A reconstruction of the Pre-Cambrian *Spriggina*. *Zoologische Jahrbuch, Anatomie* **105**, 237–258.

Birket-Smith, S. J. R. 1981b. Is *Praecambridium* a juvenile *Spriggina*? *Zoologische Jahrbuch, Anatomie* **106**, 233–235.

Bottjer, D.J., Elter, W., Hagadorn, J.W. and Tang, C.M. (eds.). 2002. *Exceptional fossil preservation*. Columbia University Press, New York, xiv+ 403pp.

Conway Morris, S. 1990. Late Precambrian and Cambrian soft-bodied faunas. *Annual Reviews of Earth and Planetary Science* **18**, 101–122.

Dewell, R. A., Dewell, W. C. and McKinney, F. K. 2001. Diversification of the Metazoa: ediacarans, colonies, and the origin of eumetazoan complexity by nested modularity. *Historical Biology* **15**, 193–218.

Erwin, D. H. 2001. Metazoan origins and early evolution. 25–31. *In* Briggs, D. E. G. and Crowther, P. R. (eds.). *Palaeobiology II*. Blackwell Scientific Publications, Oxford, xv + 583 pp.

Fedonkin, M. A. 1990. Precambrian metazoans. 17–24. *In* Briggs, D. E. G. and Crowther, P. R. (eds.). *Palaeobiology: a synthesis*. Blackwell Scientific Publications, Oxford, xiii + 583 pp.

Ford, T. D. 1958. Pre-Cambrian fossils from Charnwood Forest. *Proceedings of the Yorkshire Geological Society* **31**, 211–217.

Gehling, J. G. 1987. Earliest known echinoderm – a new Ediacaran fossil from the Pound Supergroup of South Australia. *Alcheringa* **11**, 337–345.

Gehling, J. G. 1988. A cnidarian of actinian-grade from the Ediacaran Pound Supergroup, South Australia. *Alcheringa* **12**, 299–314.

Gehling, J. G. 1991. The case for the Ediacaran fossil roots to the metazoan tree. *Geological Society of India Memoir* **20**, 181–224.

Gehling, J. G. 1999. Microbial mats in terminal Proterozoic siliciclastics: Ediacaran death masks. *Palaios* **14**, 40–57.

Glaessner, M. F. 1961. Pre-Cambrian animals. *Scientific American* **204**, 72–78.

Glaessner, M. F. 1984. *The dawn of animal life. A biohistorical study*. Cambridge University Press, Cambridge, 244 pp.

Glaessner, M. F. and Wade, M. 1966. The Late Precambrian fossils from Ediacara, South Australia. *Palaeontology* **9**, 599–628.

McMenamin, A. S. 1998. *The Garden of Ediacara*. Columbia University Press, New York, xvi + 295 pp.

Retallack, G. J. 1994. Were the Ediacaran fossils lichens? *Paleobiology* **20**, 523–544.

Runnegar, B. 1992. Evolution of the earliest animals. 65–93. *In* Schopf, J. W. (ed.). *Major events in the history of life*. Jones and Bartlett, Boston, MA, xv + 190 pp.

Schopf, K. M. and Baumiller, T. K. 1998. A biomechanical approach to Ediacaran hypotheses: how to weed the Garden of Ediacara. *Lethaia* **31**, 89–97.

Seilacher, A. 1989. Vendozoa: organismic construction in the Proterozoic biosphere. *Lethaia* **22**, 229–239.

Seilacher, A. 1992. Vendobionta and Psammocorallia: lost constructions of Precambrian evolution. *Journal of the Geological Society of London* **149**, 607–613.

Sprigg, R. C. 1947. Early Cambrian (?) jellyfishes from the Flinders Ranges, South Australia. *Transactions of the Royal Society of South Australia* **71**, 212–224.

Sprigg, R. C. 1949. Early Cambrian 'jellyfishes' of Ediacara, South Australia and Mount John, Kimberley District, Western Australia. *Transactions of the Royal Society of South Australia* **73**, 72–99.

THE BURGESS SHALE

Chapter Two

Background: the Cambrian Explosion

Although multicellular animals had only appeared at the very end of the Precambrian (Chapter 1), their evolution at the beginning of the Cambrian was so rapid that this event is known as the 'Cambrian Explosion'. In an astonishing orgy of evolution, a period of little more than 10 million years at the beginning of the Cambrian saw the appearance of almost every animal phylum and body plan known today, along with some other bizarre forms which soon became extinct, suggesting that this was an experimental phase of evolution. Approximately 35 different animal phyla are known today; in the Cambrian seas there were undoubtedly several more and some authorities would claim up to 100.

The sudden appearance of this diverse fauna within the geological record has long posed perplexing questions. Darwin supposed that these phyla had been gradually evolving throughout the Precambrian, but had simply not been preserved as they were entirely soft-bodied. Perhaps they simultaneously acquired preservable hard shells or skeletons at the onset of the Cambrian (in response to a critical change in atmospheric oxygen or oceanic chemistry) so that the Cambrian Explosion was no more than an artefact of preservation?

However, the discovery of the Precambrian Ediacaran biota (Chapter 1) in the latter half of the twentieth century showed that while soft-bodied animals did exist at the very end of the Precambrian, these were mostly primitive annelids, arthropods and cnidarians and were unlikely ancestors for the characteristic Cambrian animals such as archaeocyathids, brachiopods and molluscs. Moreover, even soft-bodied animals should leave trace fossils, which are also absent from all but the latest Precambrian sediments.

The first wave of the evolutionary explosion is evident in the basal Cambrian Tommotian Stage with the sudden appearance of many small shelly fossils. It may be, as discussed by Fortey (1997) and Conway Morris (1998), that these did have soft-bodied ancestors with a long Precambrian history, but which were so small that neither their bodies nor their traces would be preserved. Even so it is unlikely that such tiny animals, often only a few millimetres in length, were equipped with the huge range of body plans that were evident later in the Cambrian, and perhaps these 'small shellies', as they are known, are simply the hard spikes or spines of soft-skinned animals.

Very soon after their appearance the main burst of evolution began and a variety of possible triggers has been put forward. Gradually increasing oxygen levels after 3 billion years of photosynthesis by cyanobacteria and later by plants may have allowed the evolution of more mobile and larger, complex animals which were able to exploit empty niches significantly devoid of competitors. Alternatively, the emergence of the first predators may have initiated the first 'arms race', in which animals had either to escape by moving faster and/or getting bigger or to defend themselves by developing hard shells. Changes in continental configuration (and therefore of ocean currents) have also been suggested. Or maybe genetic mechanisms were simply more flexible at that time, leading to accelerated diversification.

Much of our knowledge of the fauna and flora during the Cambrian Explosion (called 'the crucible of creation' by Conway Morris, 1998) is gleaned from perhaps the best known of all Fossil-Lagerstätten – the Burgess Shale of British Columbia (Canada). By an accident of geological history, this thin layer of shale has provided a window into the richness of a Middle Cambrian marine ecosystem at a most vital time in the evolution of life on Earth. Here, by a process still not completely understood, decay was arrested so that the complete diversity of the Cambrian seas, including many soft-bodied animals (with their internal organs and muscles) has been exquisitely preserved.

These are the 'weird wonders' popularized by Stephen Jay Gould (1989) in his book *Wonderful Life*. And if one considers that approximately 85% of Burgess Shale genera are entirely soft-bodied and thus absent from other Cambrian assemblages, it becomes apparent just how misleading the fossil record would be had this particular Lagerstätte not been preserved or discovered.

History of discovery of the Burgess Shale

It was an American, Charles Doolittle Walcott, then Secretary of the Smithsonian Institution in Washington DC, who first discovered the Burgess Shale, high in the Canadian Rockies (**19**). The romantic story suggests that at the end of the 1909 field season his wife's horse, descending the steep Packhorse Trail that leads off the ridge between Mount Wapta and Mount Field, in what is now the Yoho National Park, stumbled on a boulder. Walcott dismounted to clear the track and on splitting the offending boulder revealed a fine specimen of the soft-bodied 'lace crab', *Marrella*, glistening as a silvery film on the black shale.

Unfortunately his diary does not support this story, but certainly Walcott discovered the first fossils in September 1909 and began excavating in earnest

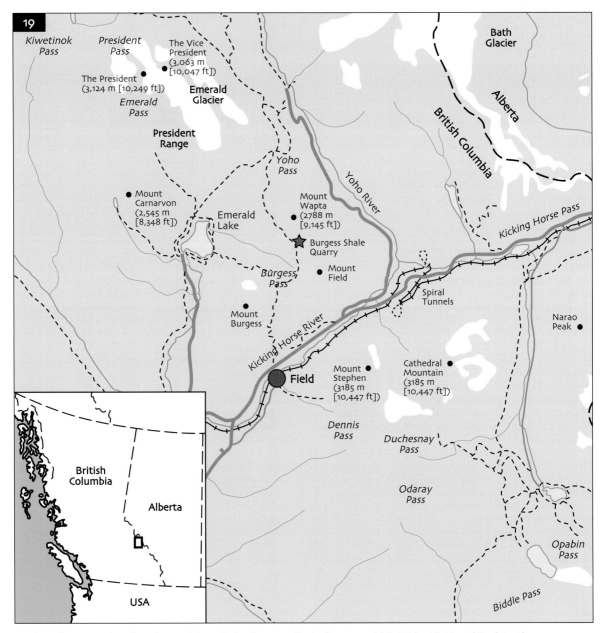

19 Locality map showing the position of the Burgess Shale Quarry within Yoho National Park in the Canadian Rockies.

during the summer of 1910. Annual field seasons with his family continued until 1913 and he returned in 1917, 1919 and 1924. By the time of his death in 1927 he had amassed a collection of 65,000 specimens, which were transported back to Washington D.C. where they remain today in the Smithsonian Institution.

By 1912 Walcott had published many of his finds and, of the 170 species currently recognized from the Burgess Shale, over 100 were described by Walcott himself.

Walcott's Burgess Shale Quarry (**20**, **21**) lies at 2,300 m (7,500 ft), just below the top of Fossil Ridge, which connects Mount Wapta and Mount Field. It is a stunning location. Looking west from the quarry, which remains snow-filled even in the summer, a breath-taking panorama of snow-capped mountains, glaciers, lakes and forests is revealed. Mount Burgess points skywards, the vivid green Emerald Lake is to its right with the Emerald Glacier poised above it. The site of Walcott's camp can be made out on the High Line Trail below.

Walcott also collected higher up the mountain at a site now called the Raymond Quarry after its detailed excavation in the 1930s by Professor Percy E. Raymond of Harvard University, whose collection now resides at the Museum of Comparative Zoology at Harvard. Apart from the work of Alberto Simonetta, an Italian biologist, who made detailed descriptions of some of the Burgess trilobites in the 1960s, little further work on these remarkable fossils was carried out until the notion of a complete restudy of the Burgess fauna was proposed by Professor Harry Whittington in 1966.

Whittington is an Englishman, but in 1966 was Professor of Palaeontology at Harvard University. He managed to persuade the Geological Survey of Canada that a restudy was timely and collecting trips in 1966 and 1967 produced over 10,000 new specimens, although few new taxa. In 1966 Whittington moved from Cambridge, Massachusetts, to Cambridge, England, and took the project with him. Very soon he recruited two young postgraduates to assist in the huge task of redescribing the Burgess animals, Derek Briggs taking on the arthropods and Simon Conway Morris the worms.

The Cambridge team made painstaking dissections, drawings and detailed photographs of the Burgess fossils and revealed a wealth of new data unseen by previous workers. Whittington produced a detailed restudy of Walcott's first find, the 'lace crab', *Marrella*, the most common animal from the Burgess fauna, and set new standards for the description of Burgess fossils. Briggs and Conway Morris worked respectively on two of the real enigmas of the Burgess Problematica, *Anomalocaris* and *Canadia sparsa* (later renamed *Hallucigenia*) and by remarkable detective work were able to elucidate the true affinities of these most bizarre Burgess animals.

In 1975 Desmond Collins of the Royal Ontario Museum (ROM) in Toronto obtained permission from the Park Rangers to collect loose material from the ridge. This party returned in 1981 and 1982, this time to look for new localities and demonstrated that the fossiliferous beds were actually more extensive than had been thought, with 10 or more new locations, both above and below Walcott's quarry, all along the line of the Cathedral Escarpment.

In 1981 the site was designated a World Heritage Site by UNESCO, but the work of the ROM team continued throughout the 1980s and 1990s with some remarkable finds.

Stratigraphic setting and taphonomy of the Burgess Shale

The Burgess Shale is a formal name which refers only to the thin unit containing the well-preserved, soft-bodied fossils, as exposed in Walcott's Quarry. This shale (the 'Phyllopod Bed' of Walcott) belongs to the Stephen Formation of Middle Cambrian age, approximately 520 million years old, which is about 150 m (500 ft) thick on Fossil Ridge.

20 Walcott's Burgess Shale Quarry at 2,300 m (7,500 ft) in Yoho National Park, Canada; Mount Wapta can be seen in the background.

21 Detail of laminated shale in the Burgess Shale Quarry, showing fining-upwards sequence with coarse, orange layers at the base and finer, grey layers above. (Measurements indicate centimetres below Walcott's 'Phyllopod Bed'.)

Just to the north of Walcott's Quarry the dark shales of the Stephen Formation abruptly disappear and abut against much lighter-coloured dolomites belonging to the Cathedral Formation with an almost vertical contact (**22**). The fact that the overlying Eldon Formation (and indeed the uppermost beds of the Stephen Formation) pass uninterrupted across this vertical contact shows that it is not a tectonic fault, but instead that it was an original feature of the Cambrian seabed, i.e. a near-vertical submarine cliff. It is significant that the Burgess Shale fossils are always found at the foot of this sub-marine cliff.

The conventional interpretation is that this cliff represents the margins of an algal reef, the top of which formed a shallow carbonate platform with well-lit waters free of terrigenous sediment. The Burgess animals lived on or in the mud in the deeper, darker water at the foot of the cliff. The seabed sloped away from the cliff into still deeper water which was anoxic and hostile to life.

Various pieces of evidence suggest that the Burgess animals were not preserved in the area in which they were living. The first is the lateral continuity over large distances of the thin beds of shale, with no evidence of bioturbation of the sediment by crawling or burrowing organisms. The second is the fact that the Burgess fossils are found lying at all possible angles within the shale, some even head-first into the sediment. Finally, examination of the thin shale beds in Walcott's Quarry reveals that each shows a definite fining-upwards sequence with coarse, orange layers at the base and finer, dark grey layers above, such that each bed represents a separate event, i.e. a separate influx of sediment (**21**).

The accepted theory of the deposition of the Burgess animals is that from time to time storms, earth movements or simply instability of the wet sediment pile sent the mud at the foot of the cliff down into the hostile basin in a rapid cloud of sediment, carrying with it the unsuspecting animals which had no time to escape. Conway Morris (1986, text-fig. 1) illustrated a 'pre-slide' and 'post-slide' environment, the former representing the area in which the animals were living, and the latter the area to which they were transported and preserved (now represented by Walcott's quarry). Both appear to have been close to the foot of the cliff, suggesting that the turbidity currents flowed downslope

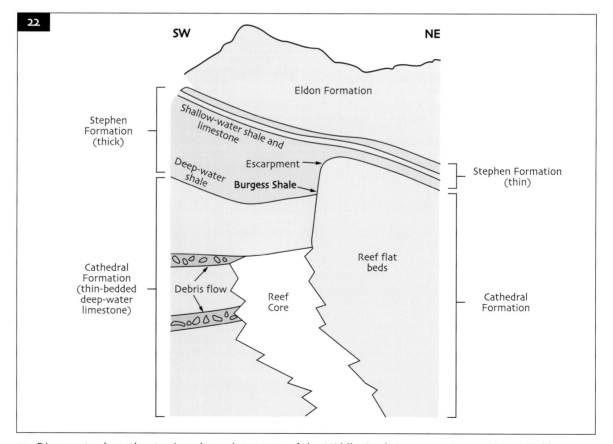

22 Diagram to show the stratigraphy and structure of the Middle Cambrian succession on Mount Field (after Briggs *et al.*, 1994).

parallel to the cliff. The slide was followed by quiet conditions allowing the fine sediment to settle, giving rise to the fine layering of the shale seen today.

How the Burgess Shale animals have been preserved is still not entirely known. Two of the common pre-requisites for soft-tissue preservation were undoubtedly partly responsible, namely rapid, catastrophic burial in a fine sediment, and deposition on a sea floor deficient in oxygen. Such a toxic 'post-slide' environment would have excluded scavengers, and when the cloud of sediment settled the carcasses would have been completely entombed with their body cavities infilled by mud. However, anaerobic microbes can break down soft muscle tissue relatively quickly even in the absence of oxygen, and some other factor must have prevented such microbial action.

Butterfield (1995) isolated tissues of various soft-bodied Burgess animals and showed that in many cases they were composed of altered original organic carbon. This was coated, however, by a thin film of calcium aluminosilicate, similar to mica, explaining the silvery appearance of the Burgess fauna under incident light.

He suggested that this coating originated from the clay minerals in the mud in which the animals were buried, and that these minerals inhibited bacterial decay – perhaps by preventing reactions of enzymes. This type of preservation is exceptional; normally very soft tissue such as muscle and intestine can only be preserved if it is replaced by another mineral during early diagenesis.

Description of the Burgess Shale biota

Vauxia (Phylum Porifera). A bush-like, branching sponge (**23**), which does not have discrete spicules, but is composed of a tough spongin-like framework, explaining why it is the most common of the Burgess sponges.

Thaumaptilon (Phylum Cnidaria; Order Pennatula-cea). A rare constituent of the sessile Burgess fauna, this possible sea pen (**24**) is nonetheless important as it is a survivor of the Precambrian Ediacara fauna (Chapter 1), resembling the genus *Charnia*. It has a broad central axis with up to 40 branches, each housing hundreds of individual star-like polyps (see Conway Morris, 1993).

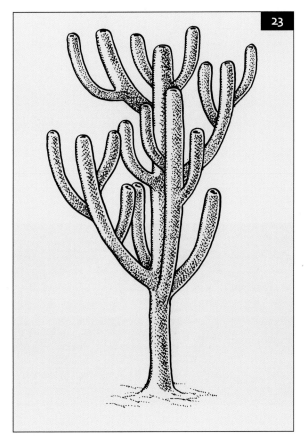

23 Reconstruction of *Vauxia*.

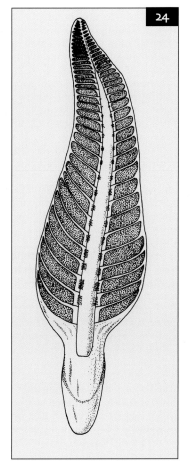

24 Reconstruction of *Thaumaptilon*.

Ottoia (Phylum Priapulida). This is the most abundant of the mud-dwelling priapulid worms (**25**, **26**), especially in the higher beds of the Burgess Shale. Priapulids are carnivorous animals, rare today, but common in the Cambrian seas. They had a bulbous anterior proboscis surrounded by vicious hooks and spines. At the end of the proboscis is a mouth with sharp teeth; their stomachs often reveal the last meal, which may includes hyoliths, brachiopods and even other specimens of *Ottoia*, for these animals were cannibals. They are commonly fossilized in a U-shape, suggesting that they lived in U-shaped burrows. However, recently discovered straight specimens may suggest that the U-shape was due to post-mortem contraction (see Conway Morris, 1977a).

Burgessochaeta and *Canadia* (Phylum Annelida; Class Polychaeta). These are polychaete worms, or bristle worms (**27**), segmented animals with paired appendages each bearing numerous bristles or setae (see Conway Morris, 1979).

Hallucigenia (Phylum Onychophora). The most celebrated Burgess animal and the classic 'weird wonder' of Stephen Jay Gould (1989) was placed in this new genus by Conway Morris to reflect its dream-like quality as it seemed unlike any other known animal (**28**, **29**). This was partly because it had been reconstructed upside-down, standing on rigid spines and waving its tentacles in the water. Additional specimens suggested reversing this interpretation and suddenly it was apparent that this genus was a marine velvet worm, a caterpillar-like group called the lobopodians ('lobed feet'). *Hallucigenia* crawled on its fleshy limbs and used its spines for protection as it scavenged for decaying food. *Aysheaia* is another such Burgess lobopod (see Conway Morris, 1977b).

Marrella (Phylum Arthropoda). A small, feathery arthropod, whose name means the 'lace-crab', it is the most common Burgess animal, with over 15,000 specimens discovered, yet it is known from no other Cambrian deposit (**30**, **31**). It was the first to be discovered by Walcott and the first to be redescribed by Whittington. A head shield has two pairs of curving spines, while the head has two pairs of antennae. The 20 body segments each bear a pair of identical legs, suggesting that it is a primitive arthropod and could be ancestral to the three major groups of aquatic arthropods (crustaceans, trilobites, chelicerates; see Whittington, 1971).

Sanctacaris (Phylum Arthropoda, Class Chelicerata). This is the most important specimen to be discovered by Collins and the ROM team as it represents the earliest known example of a chelicerate, the group containing the spiders and scorpions (**32**). The large head-shield protecting six head appendages, five of which were spiny claws to assist in capturing prey, gave it the nickname of 'Santa Claws'! (see Briggs and Collins, 1988).

Anomalocaris (Problematica or Phylum Arthropoda?). The monster predator of the Burgess Shale, *Anomalocaris* does not resemble any known animal and has long been considered as an example of a short-lived experimental arthropod-like phylum (**33**). The prey-grasping anterior appendages were originally thought to represent the segmented abdomen of a crustacean

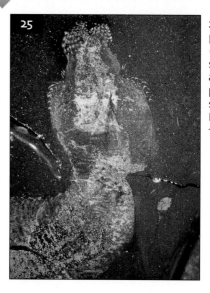

25 The priapulid worm *Ottoia prolifica*, showing anterior proboscis with spines (SI). Diameter 10 mm (0.4 in).

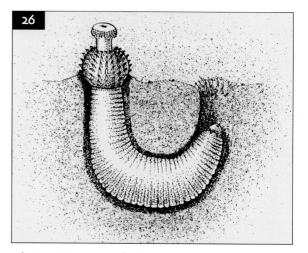

26 Reconstruction of *Ottoia*.

27 The polychaete worm *Canadia spinosa* (SI). Length about 30 mm (1.2 in).

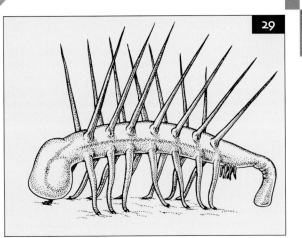

28 The velvet worm *Hallucigenia sparsa* (SI). Length about 20 mm (0.8 in).

29 Reconstruction of *Hallucigenia*.

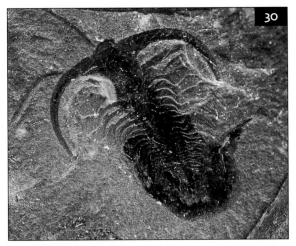

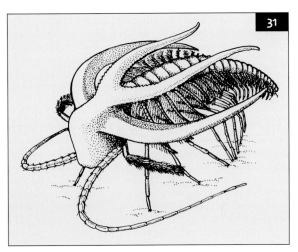

30 The arthropod *Marrella splendens* (MM). Length 20 mm (0.8 in).

31 Reconstruction of *Marrella*.

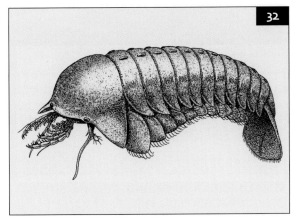

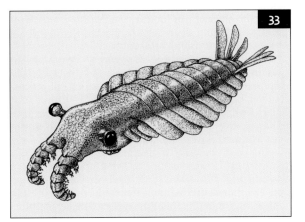

32 Reconstruction of *Sanctacaris*.

33 Reconstruction of *Anomalocaris*.

(hence the generic name meaning 'strange crab'), then were interpreted as a set of paired limbs of a giant arthropod, while the circular mouth parts were originally interpreted as the jellyfish *Peytoia*. More complete specimens revealed this to be the largest known of the Burgess animals, reaching lengths of up to 1 m (3 ft). The head has a pair of large eyes, the trunk is covered with flap-like structures, and the tail is a spectacular fan (see Whittington and Briggs, 1985).

Opabinia (Problematica). This truly strange animal had five eyes on the top of its head and a long, flexible proboscis which terminated in a number of spines, apparently a grasping organ (**34**, **35**). Each of the body segments possesses lateral lobes with gills and a strange tail was formed by three posterior flaps. Some authorities now believe that it may be related to *Anomalocaris* and that both may be arthropods (see Whittington, 1975).

Wiwaxia (Problematica). This strange mud-crawling animal was protected from predators by its dorsal surface being covered by a coat of scale-like sclerites and a double row of pointed spines (**36**). The ventral surface was a soft foot similar to that of a slug or snail and from the open mouth protruded a radula, also reminiscent of molluscs. The microstructure of the sclerites, however, is more akin to that of polychaete annelids. The true affinities of *Wiwaxia* thus remain in doubt, but it is almost certainly related to the halkieriids (**38**) (see section on Sirius Passet below and see Conway Morris, 1985).

Pikaia (Phylum Chordata). An inconspicuous but vital element of the Burgess fauna, *Pikaia* possesses a stiff rod along its dorsal margin (**37**), which suggests that it is a primitive chordate, the phylum to which Man and all vertebrates belong, and showing that even our ancestors were present during the Cambrian Explosion. A narrow anterior end has a pair of tentacles, while the posterior is expanded into a fin-like tail. The cone-in-cone arrangement of the muscles is often clearly preserved.

Palaeoecology of the Burgess Shale

The Burgess Shale represents a marine, benthic community living in, on, or just above the muddy seabed at the foot of a submarine cliff, where the mud was banked sufficiently high to be clear of stagnant bottom waters. The marine basin faced the open sea and was situated within the tropical zone at about 15°N. The presence of photosynthesizing algae suggests that the depth was not much more than 100 m (300 ft).

Palaeoecological analysis by Conway Morris (1986) examined over 30,000 slabs of shale with 65,000 fossils. Approximately 10% of the biota consisted of benthic infauna, i.e. living in the sediment itself, this community being dominated by the burrowing priapulid worms such as *Ottoia*, *Selkirkia* and *Louisella*, and polychaete worms such as *Burgessochaeta* and *Canadia*.

The vast majority of Burgess animals (c. 75%) consisted of benthic epifauna, living on the sediment surface, and divided into the fixed or sessile epifauna (c. 30%) and the vagrant epifauna (c. 40%), that walked or crawled across the seabed. The sessile biota consisted mainly of sponges, such as the branching *Vauxia*, and others, such as *Choia* and *Pirania* adorned with

34 The problematical *Opabinia regalis* (SI). Length about 65 mm (2.5 in).

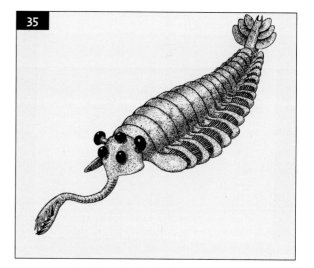

35 Reconstruction of *Opabinia*.

36 The problematical *Wiwaxia corrugata* (SI). Length about 35 mm (1.4 in).

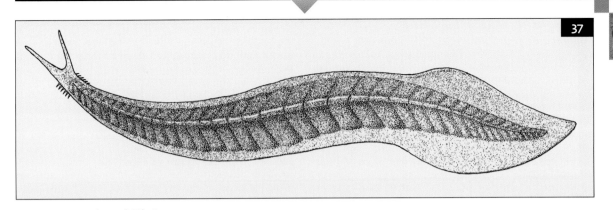

37 Reconstruction of *Pikaia*.

38 The halkieriid *Halkieria evangelista*, from Greenland (GMC). Length 60 mm (2.4 in).

sharp, glassy spicules which both supported the skeleton and protected against predators. The sea pen *Thaumaptilon* was a rare member of the sessile epifauna, as was the enigmatic animal *Dinomischus*, projecting just 1 cm (0.4 in) above the mud on a thin stalk and looking like a small flower.

The vagrant epifauna was more varied, but was dominated by arthropods, only a small proportion of which were trilobites, such as *Olenoides* and the soft-bodied *Naraoia*. It also included the ubiquitous 'lace crab', *Marrella*, and a number of small, scavenging lobopods, most notably *Hallucigenia* and *Aysheaia*. The most bizarre mud crawler was *Wiwaxia*, which moved across the sediment using its slug-like foot.

Animals which lived above the mud surface were fewer in number, comprising only about 10% of the Burgess biota, simply because they were more able to escape the mud flow by swimming away. The nekto-benthic animals (near-bottom swimmers) were, however, within the range of turbidity flows and included the tiny chordate *Pikaia*. The medusoid *Eldonia*, which is more probably related to the holothurians (sea cucumbers) than to true jellyfish, may be regarded as nektobenthos and/or a planktonic floater. Nekton (forms swimming above the substrate) include the giant

predator *Anomalocaris* and the enigmatic *Opabinia*, which used its nozzle-like proboscis to search for food.

Trophic analysis by Conway Morris (1986) identified several feeding types including: filter feeders, dominated by the sponges; deposit feeders, dominated by arthropods (deposit collectors) and molluscs (deposit swallowers); scavengers such as the lobopods *Hallucigenia* and *Aysheaia*; and lastly, predators such as the giant *Anomalocaris*, large arthropods (such as *Sidneyia*) and the priapulids. The important role of predation evident in the Burgess Shale ecological framework provides the main contrast with other Cambrian (shelly) assemblages.

Comparison of the Burgess Shale with other Cambrian biotas

Burgess Shale-type assemblages have since been discovered at about 40 localities worldwide, but two are of special significance.

Sirius Passet, northern Greenland

Discovered in 1984 by the Geological Survey of Greenland near J. P. Koch Fjord in Peary Land, north Greenland, this locality, now known as Sirius Passet, was first seriously collected by a 1989 expedition. It soon became apparent that this was another predominantly soft-bodied Cambrian fauna, also seemingly deposited in deep water muds adjacent to a shallow carbonate bank, and more than 3,000 specimens were collected from what is now termed the Buen Formation. Significantly, however, this fauna is of Lower Cambrian age (Atdabanian), somewhat older than the Burgess Shale, revealing that the Cambrian Explosion was well under way by this time.

One of the first and most intriguing specimens to be discovered at Sirius Passet was *Halkieria*, a slug-like animal (**38**) with a dorsal coat of scale-like sclerites just as in *Wiwaxia* (**36**). But, at either end of the long body is a shell looking remarkably like an inarticulate brachiopod (see Conway Morris and Peel, 1990; Conway Morris, 1998, fig. 86). Perhaps these animals are telling us that molluscs, annelids *and* brachiopods are phylogenetically closer than had been supposed? Molecular biology supports this conclusion.

Chengjiang, southern China

'Discovered' in the same year, 1984 (although actually known since 1912), the Chengjiang Fossil-Lagerstätte, best exposed at Maotianshan, near Kunming, in Yunnan Province, southern China, also yields abundant soft-bodied 'Burgess Shale-type' fossils (see Chen and Zhou, 1997; Hou *et al.*, 2004)). Although of similar age to the Sirius Passet assemblage (Lower Cambrian, Atdabanian Stage), many of the characteristic Burgess animals are known from Chengjiang (including complete specimens of *Hallucigenia* and *Anomalocaris*), in addition to new Chinese genera of arthropods, worms, sponges, brachiopods and so on. The faunal similarity is remarkable because the South China craton would have been thousands of kilometres from the continent of Laurentia (comprising North America and Greenland).

Amongst several new groups, the new phylum,

Vetulicolia, was proposed by Shu *et al.* (2001) to include segmented, arthropod-like metazoans which had obvious gill slits, suggesting a deuterostome affinity. But perhaps the most surprising element of the fauna was reported by Shu *et al.* (1999): the discovery of agnathan fish (previously known from the Lower Ordovician), illustrating that even the vertebrates appeared during the Cambrian Explosion.

Fossil preservation is spectacular with reddish-purple impressions on fine orange shale, and again appears to be the result of rapid, live burial in catastrophic turbidity flows. The Chengjiang animals probably lived in the area of their burial, which was adjacent to a delta front. The 50 m (160 ft) thick Maotianshan Shale Member of the Yu'anshan Formation consists of thin, graded mudstone layers, recording short episodic sedimentation events.

Further Reading

Briggs, D. E. G. and Collins, D. 1988. A Middle Cambrian chelicerate from Mount Stephen, British Columbia. *Palaeontology* **31**, 779–798.

Briggs, D. E. G., Erwin, D. H. and Collier, F. J. 1994. *The fossils of the Burgess Shale.* Smithsonian Institution Press, Washington DC, xvii + 238 pp.

Butterfield, N. J. 1995. Secular distribution of Burgess Shale-type preservation. *Lethaia* **28**, 1–13.

Chen, J. and Zhou, G. 1997. Biology of the Chengjiang fauna. *Bulletin of the National Museum of Natural Science* **10**, 11–105.

Conway Morris, S. 1977a. Fossil priapulid worms. *Special Papers in Palaeontology* **20**, 1–95.

Conway Morris, S. 1977b. A new metazoan from the Cambrian Burgess Shale of British Columbia. *Palaeontology* **20**, 623–640.

Conway Morris, S. 1979. Middle Cambrian polychaetes from the Burgess Shale of British Columbia. *Philosophical Transactions of the Royal Society of London*, Series B **285**, 227–274.

Conway Morris, S. 1985. The Middle Cambrian metazoan *Wiwaxia corrugata* (Matthew) from the Burgess Shale and *Ogygopsis* Shale, British Columbia. *Philosophical Transactions of the Royal Society of London*, Series B **307**, 507–582.

Conway Morris, S. 1986. The community structure of the Middle Cambrian Phyllopod Bed (Burgess Shale). *Palaeontology* **29**, 423–467.

Conway Morris, S. 1993. Ediacaran-like fossils in Cambrian Burgess Shale-type faunas of North America. *Palaeontology* **36**, 593–635.

Conway Morris, S. 1998. *The crucible of creation.* Oxford University Press, Oxford, xxiii + 242 pp.

Conway Morris, S. and Peel, J. S. 1990. Articulated halkieriids from the Lower Cambrian of north Grenland. *Nature* **345**, 802–805.

Fortey, R. 1997. *Life: an unauthorised biography.* Flamingo, London, xiv + 399 pp.

Gould, S. J. 1989. *Wonderful Life: the Burgess Shale and the nature of history.* Norton, New York, 323 pp.

Hou, X. G., Aldridge, R. J., Bergström, J., Siveter, D. J., Siveter, D. J. and Feng, X. H. 2004. *The Cambrian fossils of Chengjiang, China.* Blackwell, Oxford, 256pp.

Shu, D. G., Luo, H. L., Conway Morris, S., Zhang, X. L., Hu, S. X., Chen, L., Han, J., Zhu, M., Li, Y. and Chen, L. Z. 1999. Lower Cambrian vertebrates from south China. *Nature* **402**, 42–46.

Shu, D. G., Conway Morris, S., Han, J., Chen, L., Zhang, X. L., Zhang, Z. F., Liu, H. Q., Li, Y. and Liu, J. N. 2001. Primitive deuterostomes from the Chengjiang Lagerstätte (Lower Cambrian, China). *Nature* **414**, 419–424.

Whittington, H. B. 1971. Redescription of *Marrella splendens* (Trilobitoidea) from the Burgess Shale, Middle Cambrian, British Columbia. *Bulletin of the Geological Survey of Canada* **209**, 1–24.

Whittington, H. B. 1975. The enigmatic animal *Opabinia regalis*, Middle Cambrian, Burgess Shale, British Columbia. *Philosophical Transactions of the Royal Society of London*, Series B **271**, 1–43.

Whittington, H. B. and Briggs, D. E. G. 1985. The largest Cambrian animal, *Anomalocaris*, Burgess Shale, British Columbia. *Philosophical Transactions of the Royal Society of London*, Series B **309**, 569–609.

THE SOOM SHALE

Chapter Three

Background: early Palaeozoic Lagerstätten

Between the extraordinary biotas of the Lower and Middle Cambrian (e.g. Burgess Shale and Chengjiang, Chapter 2) and the exceptional early terrestrial biotas of Siluro-Devonian times, e.g. Rhynie, Ludlow, and Gilboa, described in Chapter 5, the fossil record lacks major Lagerstätten. The Ordovician Period, in particular, is fairly barren of exceptional biotas. However, one Ordovician horizon, the Soom Shale Member of the Table Mountain Group of Lower Palaeozoic sedimentary rocks of the Western Cape, South Africa, sprang to fame in the 1990s with the discovery of giant conodont animals with preserved musculature and feeding apparatuses. The Soom Shale is the most important Ordovician Lagerstätte and is unique in that it comes from a high-latitude (60°S), cool, glacially influenced, marine habitat. In evolutionary terms, some of the animals of the Soom Shale hark back to the Middle Cambrian biotas – for example, there is a naraoiid-like trilobite *Soomaspis* – yet other animals of the Soom Shale, such as the eurypterid *Onychopterella*, presage a greater diversity to come.

History of discovery of the Soom Shale

Because it is relatively soft, the Soom Shale forms a distinct bench in the landscape of rugged sandstone escarpments and plateaus which dominates the scenery of the south-western Cape region, so its existence has been well known to geologists for many years. Few fossils occur in the predominantly sandstone rocks of the Table Mountain Group, to which the Soom Shale belongs and, at first sight, the Soom Shale appears to be equally barren. Indeed, it was not until 1958 that fossil tracks and trails were discovered in the Table Mountain Group. The presence of some brachiopods and fragments of trilobites were provisionally announced in 1967, and later published by Cocks *et al.* in 1970, which gave the first indication of the Ordovician age of these shales. This was the first Lower Palaeozoic fauna to be described from South Africa.

The fossils of the Soom Shale first rose to fame in 1993 with the report of eyes in a conodont animal. It was only in 1983 that conodonts – usually small, phosphatic tooth-rows, sometimes found arranged in basket-like structures – were found in association with soft tissues which later (1993) were proved to be those of early fish. It was the evidence of cartilage supporting large eyes in the Soom Shale conodont, *Promissum pulchrum*, that helped to determine the chordate nature of conodonts. Following this discovery, a concerted effort was made to find more specimens in the Soom Shale. This is a daunting task because fossils are rare in the shale, but dedicated hunting by Dick Aldridge and Sarah Gabbott, of Leicester University, UK, and Johannes Theron, of the Geological Survey of South Africa, has since unearthed a wide variety of animals and plants. Further specimens of *Promissum* turned up, which added more information about the morphology of conodonts and, in 1995, a specimen with muscle tissue was reported. Other exciting animals from the Soom Shale include the eurypterid ('sea scorpion') *Onychopterella*, which also shows details of musculature as well as the gut, enigmatic naraoiid trilobites, and orthocone cephalopods (relatives of squids and ammonites with straight, conical shells).

Stratigraphic setting and taphonomy of the Soom Shale

The Soom Shale Member is maximally 10 m (30 ft) thick, characteristically finely laminated and yellow-brown to light or dark grey (see below). Above it lies the coarser, buff Disa Siltstone Member, about 130 m (425 ft) thick. Together, these two members comprise the Cedarberg Formation (39), named after the beautiful Cedarberg mountains (40), which form a north–south ridge between Citrusdal and Clanwilliam in the Western Cape, some 150 km (90 miles) north of Cape Town (41). The Cedarbergs are named after the endemic Clanwilliam Cedar tree, *Widdringtonia cedarbergensis*, which, after extensive logging in the eighteenth and nineteenth centuries, is now sadly represented only by gnarled and twisted specimens on mountain tops above 1,000 m (3,300 ft). The age of the Soom Shale has been determined as Ashgill (late Ordovician) on the basis of the occurrence of the trilobite *Mucronaspis olini* by Cocks and Fortey (1986) in an outcrop in the Hexrivier Mountains, some 100 km (60 miles) from the main locality at Clanwilliam. Most of the biota has come from the locality (42) at Keurbos Farm, about 13 km (8 miles) south of Clanwilliam; some animals come from the farm of Sandfontein, 25 km (15 miles) east of Citrusdal.

The Table Mountain Group is a 4,000 m (13,000 ft) thick pile of predominantly sandy sediments. Two of the lower formations, the Graafwater and Peninsula, form the distinctive Table Mountain at Cape Town, but the higher beds do not appear until further north. The Pakhuis Formation lies between the Peninsula and Cedarberg Formations and is the clue to understanding the environment of the time. The Pakhuis Formation consists predominantly of tillite – fossilized till or boulder clay, the product of glaciers. In many places beneath the Pakhuis Formation, the tillites can be seen to rest on a grooved pavement of the Peninsula Formation – glacial striae which were carved as ice moved across and eroded the Peninsula Formation rocks. Because it is an erosion surface, the base of the Pakhuis Formation is an unconformity, but its top grades gradually into the finely laminated Soom Shale. Therefore, the Soom Shale appears to represent quiet water not far from the ice front. The mud and silt laminae, which are generally in the order of 1 mm (0.04 in) thick, rarely up to 10 mm (0.4 in), and are laterally persistent, represent the gentle settling of fine particles derived from the ice meltwater. These silt layers may represent distal turbidites – the final settling-out of a flurry of sediment which spread out across the floor of the sea (or possibly brackish bay), or possibly seasonal

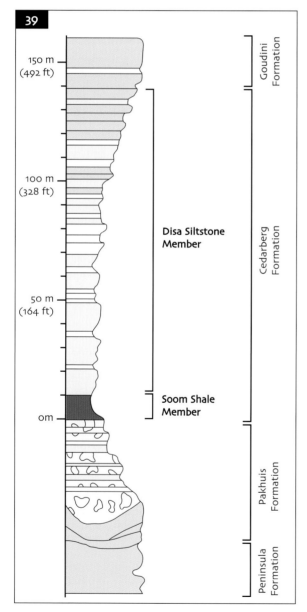

39 Stratigraphic section showing the position of the Soom Shale Member of the Cedarberg Formation, Table Mountain Group (after Theron *et al.*, 1990).

40 San rock art on the wall of a cave in Cedarberg Formation sandstone, with a view to the Cedarberg Mountains across the Brandewyn River valley.

41 Map showing the outctrop of the Table Mountain Group in the Cape region of South Africa (after Theron *et al.*, 1990).

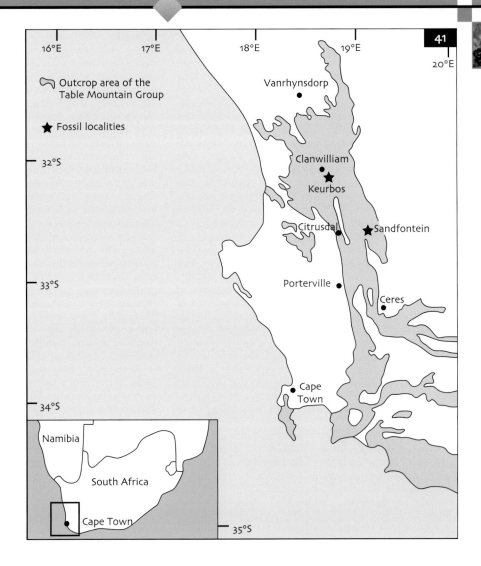

freezing and thawing (varves). There is some evidence for floating ice in the area too, in the form of dropstones – pebbles dropped from passing icebergs onto the muddy seabed as the icebergs melted. The overlying Disa Siltstone Member is composed of coarser siltstones with evidence of occasional wave and current activity; thus it represents shallow marine water with a greater input of sediment as the ice sheets melted.

The Soom Shale shows no signs of bioturbation (animal activity in the sediment), which suggests that it was not possible for animals to live within the sediment or directly on the sea floor. This may have been because the temperature was too low although, as we shall see later, there were animals living in the waters above. Perhaps the mud and water on the sea floor were anoxic (lacking oxygen) or toxic in some way. Many of the laminate surfaces have ribbon-like algae ('seaweeds') on them, though these algae were not necessarily benthic (bottom-living); they could have descended to the sea floor after dying at the surface.

42 Quarry in Soom Shale at Keurbos Farm, 13 km (8 miles) south of Clanwilliam, Western Cape.

The Soom Shale animals are preserved as thin films of clay minerals replacing the original organic and/or mineral tissues of the carcasses. Shale is highly compressible, so the fossils are quite flattened. The sequence of events which led to the preservation of the fossils is interesting, and some aspects of it are unique. From the evidence given above, we can conclude that the water was cool, shallow (or not too deep), and there were no currents. Had there been currents, we would expect to see not only some rippling of the shale surfaces perhaps, but also alignment of elongate fossils such as the enigmatic spines called *Siphonacis*. The bottom waters were probably anoxic; there are virtually no benthic animals, no bioturbation, and the animals appear to be nearly complete and not scavenged or much decayed, as would normally occur on an oxygenated sea floor. It is possible that the sea-floor waters were not always anoxic, but that biota were only preserved during times of anoxia. Cold water appears to inhibit decay, and to increase the dissolution of calcium carbonate, which helps to explain the lack of calcitic shells but the preservation of organic tissues. At an early stage, organic material was replaced by the clay mineral illite. Later, apatite, as found in conodont teeth and parts of some brachiopod shells, was replaced by silica. There has been some debate (see Gabbott, 1998; Gabbott *et al.*, 2001) as to how organic tissues can be replaced by clay minerals, and whether the original replacement was by kaolinite which later changed to illite, or whether the illite replaced the organic molecules directly. Gabbott prefers the latter process.

What makes the preservation of the Soom Shale unique is that there are no other Lagerstätten in which direct replacement of organic materials by clay minerals has been demonstrated. The Burgess Shale (Chapter 2) animals are preserved as films of shiny micaceous minerals, similar to clays, as well as some clays adhering to the cuticles of trilobites. However, these minerals appear to have been adsorbed onto the organic surfaces, and formed rather later in the diagenetic process than the clay replacement in the Soom Shale. It is possible that clay mineral adsorption preceded replacement in the Soom Shale. Further work is necessary before we can be sure whether the Soom Shale is unique in this respect, or whether it represents an end-member of a geochemical spectrum.

Description of the Soom Shale biota

Compared to other Ordovician shallow-water communities, as found for example in localities in North Wales and Scotland, the Soom Shale biota is considerably restricted in diversity. There are microfossils such as plant spores, acritarchs and chitinozoans (the last two are organic-walled cysts of unknown affinity).

Microfossils. Chitinozoa are particularly well preserved in the Soom Shale. Like acritarchs, chitinozoans are organic-walled, generally flask-shaped microfossils of uncertain affinity, but they are known only from rocks of Ordovician to early Carboniferous age, and whereas acritarchs appear (from their biochemical composition, similar to sporopollenin) to be plant-related, chitinozoans are composed of a chitin-like

substance, which suggests an animal affinity. Moreover, while chitinozoans are usually found as isolated individuals after maceration of the rock matrix, they can also be found on rock surfaces in strings, radial aggregates and clusters inside a membrane suggestive of a cocoon. Search of the rock surface on and around the Soom Shale conodont animals and orthocone cephalopods by Gabbott *et al.* (1998) revealed all four types of occurrence – individuals, strings, aggregates and cocoons. The most likely hypothesis of the zoological affinity of chitinozoans is that they are eggs and egg-masses or cocoons of an animal group which had the same stratigraphic range as the Chitinozoa: Ordovician to early Carboniferous. This narrows the possibilities down somewhat, and then if it is assumed that the producer of the chitinozoans is to be found among the restricted fauna of the Soom Shale, the possibilities are reduced to conodont animals or ortho-cone cephalopods. Other possible contenders, such as graptolites and gastropods, can be ruled out because they do not occur, or are found only rarely, in the Soom Shale. There is more association between chitinozoans and cephalopods than conodonts in the Soom Shale occurrences, so it is slightly more likely that Chitinozoa represent egg-masses of orthocone cephalopods.

Brachiopods. Few macrofossils are abundant in the Soom Shale, but some brachiopods are found more commonly than other fossils. These include orbiculoid and lingulate inarticulates (with phosphatic shells), and a few ribbed articulates (with calcified shells). *Lingula* is only found in shallow water and is infaunal, so needs an oxygenated substrate; its presence suggests that the sea floor during Soom Shale times was at least occasionally oxygenated because there is no evidence that it drifted in from elsewhere (i.e. it is autochthonous). While over 100 orbiculoid brachiopods have now been collected scattered in the shale at the Keurbos locality, vastly more have been found in association with orthocone cephalopods. The orbiculoids occur attached as epizoans to the orthocone shells and, because the brachiopods occur all over the orthocones with no preferred side or orientation, it is presumed that they colonized the orthocones when they were alive and swimming in the nekton. Furthermore, if the brachiopods colonized the orthocone in life, then one would expect them to be larger nearer the apex of the shell (the end which formed earliest) than the aperture (which is the part of the shell where growth continues). One particularly large and well-preserved orthocone specimen and its associated epizoa was studied by Gabbott (1999) (**43**). She found that while there was no great difference in the size of brachiopods from the apex to the aperture of the orthocone, there were fewer brachiopods near the apex, and the loose brachiopods found in the nearby rock matrix were generally larger than those on the orthocone. To explain this distribution it is necessary to understand what might happen to an orthocone when it died. The carcass would fall to the sea floor and begin to decay. Buoyancy gases in the shell, perhaps combined with gases produced by decay, would buoy up the shell apex while the aperture would become buried in the sediment, so the brachiopods near the aperture would be covered up and die, while those (larger ones) near the apex might well also die

43 Orthocone nautiloid with associated orbiculoid brachiopod epizoans (GSSA). Length 243 mm (9.5 in).

44 The trilobite *Soomaspis splendida* (GSSA). Length 30 mm (1.2 in).

and drop off the exposed shell since they were now in a poorly oxygenated environment.

Trilobites. The widespread Upper Ordovician trilobite genus *Mucronaspis* has been identified in the Soom Shale, some 100 km (60 miles) away from the main Keurbos locality, and it is this trilobite, together with the brachiopods, which provided the dating of the Soom Shale. Of greater interest is the strange animal *Soomaspis splendida* (**44**), which was described by Fortey and Theron (1995). This animal resembles an agnostid trilobite in having no eyes, just three thoracic segments, and cephalon and pygidium about the same size (isopygous). However, its uncalcified cuticle places it with the naraoiids (see *Naraoia* from The Burgess Shale, Chapter 2). The discovery of *Soomaspis* prompted Fortey and Theron to re-examine the relationship between the naraoiids and the rest of the trilobites. Because of the reduced number of thoracic segments, agnostid and naraoiid trilobites superficially resemble young stages of more typical trilobites, which begin life without segments and add them one by one through ontogeny. In addition, the small adult size and reduced number of appendages of agnostids suggests that they attained their state by early onset of maturation in a

juvenile form of their ancestors – an evolutionary process termed progenesis. The evolutionary benefit this gave the agnostids was to enable them to bypass a long existence as part of the benthos and instead enjoy a rapid turnover of generations entirely within the plankton. Naraoiids, on the other hand, had few or no thoracic segments, were larger trilobites and, as evidenced from *Naraoia* from the Burgess Shale, had a considerable number of appendages beneath their uncalcified shields. Thus, to produce a naraoiid only the production of thoracic segments is inhibited, while growth rate is normal or enhanced and maturation occurs at usual adult size. This process is termed hypermorphosis. Because of these different developmental routes, the small size of agnostids and their calcified exoskeleton, Fortey and Theron (1995) argued that naraoiids and agnostids were unrelated to each other, though possibly both could still be accommodated in the Trilobita. While agnostids can be derived from normal Cambrian trilobites, and their small size, loss of eyes, and so on are secondary, naraoiids appear to be primitive trilobites which never developed dorsal eyes, a calcified cuticle, or the numerous segments of the typical trilobite.

Eurypterid. A few myodocopid ostracodes have been found in association with the orthocone cephalopods, but the most interesting other arthropod in the Soom Shale is the eurypterid *Onychopterella* described by Braddy *et al.* (1995) (**45**). Eurypterids ranged from the Ordovician to the Permian periods, with their acme in the Silurian. Early eurypterids were marine animals, but by the late Silurian they had apparently invaded brackish and fresh waters, even becoming amphibious. They ranged up to 2 m (6 ft) in length (there is evidence of a giant eurypterid trackway made by an animal of about this size in Permian sediments near Laingsburg just 150 km (90 miles) south-east of the Cedarberg range), and were thus the largest arthropods which ever lived. They were the top predators on Earth for some 200 million years. However, of the two Soom Shale eurypterid specimens, the larger would have measured less than 150 mm (6 in) in life. Several features make the Soom eurypterid important. First, that few southern hemisphere eurypterids are known, and *Onychopterella* was previously only recorded from the Silurian of North America. Therefore the Soom occurrence extends the genus into the Ordovician, and also the family Erieopteridae, to which it belongs, into the former Gondwanaland. Second, the eurypterid has a peculiar

45 The eurypterid *Onychopterella* (GSSA). Length 150 mm (6 in).

spiral structure preserved between the giant coxae of the last pair of appendages (the swimming legs). This structure was interpreted by Braddy *et al.* (1995) as a spiral valve in the gut, a feature normally associated with animals which scavenge on detritus in mud, but also known from another South African eurypterid *Cyrtoctenus.* Clearly, since nothing lived on the sea floor in Soom times, it is presumed that *Onychopterella* was not a mud-grubber; moreover, the spiral valve found in fish and *Cyrtoctenus* is further back in the alimentary system, so its presence in *Onychopterella* is problematical. Third, *Onychopterella* also has gill lamellae preserved. It has been known for a long time that eurypterids used gills for respiration in water, but all that had ever been preserved was a rather characteristic spongy material whose surface area for gas exchange is too small to support the respiratory needs of such large animals. Selden (1985) discussed this palaeophysiological enigma, concluding that the spongy material was actually an accessory organ for breathing air when on land, such as can be found in amphibious crabs and woodlice, and predicted that true gill lamellae would be thin and only preserved in exceptional circumstances. The Soom Shale has provided the exceptional preservation required to see these essential organs (see Braddy *et al.*, 1999).

Orthocone cephalopods. Fourteen orthocone cephalopods from the Keurbos farm locality were described by Gabbott (1999) (**43**). All specimens preserve the body chamber and some also retain the phragmocone (the rest of the shell, which contains gas chambers). The largest body chamber is 103 mm (4 in) long, and the longest phragmocone is 243 mm (9.5 in) long and 44 mm (1.7 in) at its widest point, so the largest specimen would have been just less than 350 mm (14 in) long. All are greatly flattened and preserved only as moulds because the original aragonite has dissolved. The adhering epizoans have already been described, but there are other features of the orthocones which are exceptionally preserved. Radulae, rows of teeth which the orthocone used for feeding, are preserved as external moulds of teeth. Radulae are rarely preserved in fossil cephalopods, and are more commonly found in Mesozoic ammonites than Palaeozoic cephalopods. Indeed, radulae have only so far been reported in Palaeozoic non-ammonoids from the Upper Carboniferous Mazon Creek of Illinois (Chapter 6) and the Silurian of Bolivia. So the Soom Shale radulae are the oldest so far recorded for this group. There are two types of radulae known from cephalopods: a row of thirteen elements in nautiloids and a row of seven in coleoids (squids and cuttlefish) and ammonoids. The Mazon Creek nautiloid shows thirteen elements, so is presumed to be a nautiloid, and the Silurian orthoceratid from Bolivia appears to show seven, suggesting a closer relationship with coleoids and ammonoids than nautiloids. Only four radular elements can be seen in the Soom Shale cephalopod, and it is possible that there were more elements which were not preserved or are hidden by the four visible elements, so the Soom Shale material does not help in understanding the evolution of cephalopod radulae.

Conodonts. The Soom Shale is perhaps best known for its exceptional preservation of conodont animals (**46**). From their discovery well over a century ago until just 20 years ago, the animal to which the simple to complex, spiny, phosphatic structures known as conodonts belonged remained tantalizingly obscure. A number of, mostly worm-like, contenders had been put forward but in most cases it was difficult to prove that the enclosing soft-bodied creature had not simply eaten a conodont animal. In 1982 a fossil was discovered in which it could be shown that the conodont elements were truly part of the anatomy of the animal (Briggs *et al.*, 1983), and that animal was a chordate, perhaps related to the primitive jawless amphioxus and hagfish.

The Soom Shale conodont animal, *Promissum pulchrum*, was first described in 1986 as a very early land plant which could hold an important position in the evolution of land plants (the name translates as 'beautiful promise') (Kovács-Endrödy, 1987). The reason that *Promissum* was mistaken for a plant is that, while conodonts are most commonly found as isolated elements after acid-digestion of rocks, when they are large enough to be seen on bedding planes, conodont elements are sometimes found arranged into apparatuses which are thought to represent feeding structures. While it was obvious to some who studied the specimens (e.g. Rayner, 1986) that the fossils were not plants, it was not easy to say what they really were: giant graptolites perhaps? The graptolite expert Barrie Rickards studied the specimens and eventually concluded that they could be large conodonts, and so brought in Dick Aldridge, conodont expert, to confirm their identity. Thus, *Promissum pulchrum* became known as the largest conodont assemblage known, and clearly belonged to a giant animal (Theron *et al.*, 1990). An intensive search for more specimens followed and, in 1993, the hunt was rewarded when preserved soft tissues of *Promissum pulchrum* were found (Aldridge and Theron, 1993). Later finds from the locality at Sandfontein gave more information about the type of animal which bore the conodont elements. It was up to 400 mm (16 in) long, with a pair of eyes at the front end, a notochord (stiff rod) along its trunk, and myomeres (muscle blocks), which suggests that it propelled itself with an eel-like wave passing down its trunk and tail. The conodont elements are positioned behind and below

the eyes, and were clearly involved in feeding. However, one aspect of conodont biology which took some time to be resolved concerned how conodont apparatuses functioned as a feeding mechanism. Broadly, one school of thought contended that the elements were used as teeth or raking devices, while the other (pointing to lack of wear and the growth pattern) maintained that they were internal supports for a fleshy structure. Chemical analysis showed that conodont apparatuses are made of the same type of bone and enamel as in modern vertebrates, and functional morphological studies have shown how they were used for feeding, yet their complex, diverse structure is wholly unlike that of any primitive fish, so their feeding method could not have given rise directly to the earliest fish jaws.

Miscellanea. Like most other Fossil-Lagerstätten in this book, the Soom Shale contains some enigmatic creatures which cannot be placed in known groups. The most interesting is a large (400 mm [16 in] long), multi-segmented, soft-bodied animal of unknown affinity which has yet to be fully studied and published (**47**). Also, there are scattered, organic-walled spines given the name of *Siphonacis*, which could belong to a spiny animal of some sort. Finally, among the epizoa attached to the orthocone shells are some cornulitids – Palaeozoic cone-shaped fossils of uncertain affinity.

47 Undescribed enigmatic animal (GSSA). Length 400 mm (16 in).

46 The conodont anterior apparatus *Promissum pulchrum*; eyes bottom left, conodont apparatus to right (GSSA). Length 22 mm (0.9 in).

Palaeoecology of the Soom Shale

Gathering together the evidence provided by the Soom Shale biota, and the data given by the sedimentology, a picture can be painted of a shallow sea with a muddy, cold, mainly lifeless bottom. In the water body there were, however, many swimming organisms (nekton). They occupied a number of feeding niches: the eurypterids, conodonts and orthocones were presumably predators and/or scavengers, while the brachiopods and cornulitids were filter feeders. Occasionally the sea floor allowed some benthos, such as the lingulate brachiopods (filter feeders) and scavenging myodocopid ostracodes. There is also evidence of undiscovered large predators or scavengers in the form of coprolites, crushed brachiopod shells and broken conodont elements.

Comparison of the Soom Shale with other Lower Palaeozoic biotas

As mentioned in the introduction to this chapter, there are virtually no other exceptionally preserved biotas in the Ordovician with which to compare the Soom Shale. Nevertheless, the horizon reflects an interesting period in late Ordovician times which is reflected in other, less well preserved, biotas elsewhere. The Soom Shale correlates to the late Ordovician, Ashgill Epoch, and specifically the Hirnantian Age, the last in the Ashgill. At this time, there is evidence for a glaciation centred on the part of Gondwana now occupied by North and West Africa. Its effects were more wide-ranging, however. In Britain, for example, it is noticeable that there was a severe decline in the diversity of graptolites, plankton which are normally associated with low, subtropical latitudes. Indeed, the end-Ordovician mass extinction event which, like the more famous end-Cretaceous extinction which wiped out the dinosaurs, extinguished nearly a quarter of animal families living at the time, has been linked to the Hirnantian glaciation. Three-quarters of all trilobites and a quarter of all brachiopods became extinct at the time, and were replaced by the so-called Hirnantia Fauna of animals more suited to life in cool waters. The Hirnantia Fauna can be seen in a number of localities around the world. The Silurian Period is similarly as sparse as the Ordovician in exceptional biotas, although fairly recently a new Lagerstätte from the Wenlock (lower Silurian) has been reported (Briggs *et al.*, 1996). It is preserved in volcanic ashes and its fauna is in the early stages of description.

Further Reading

Aldridge, R. J. and Theron. J. N. 1993. Conodonts with preserved soft tissue from a new Ordovician Konservat-Lagerstätte. *Journal of Micropalaeontology* **12**, 113–117.

Aldridge, R. J., Theron. J. N. and Gabbott, S. E. 1994. The Soom Shale: a unique Ordovician fossil horizon in South Africa. *Geology Today* **10**, 218–221.

Braddy, S. J., Aldridge, R. J. and Theron, J. N. 1995. A new eurypterid from the late Ordovician Table Mountain Group, South Africa. *Palaeontology* **38**, 563–581.

Braddy, S. J., Aldridge, R. J., Gabbott, S. E. and Theron, J. N. 1999. Lamellate book-gills in a late Ordovician eurypterid from the Soom Shale, South Africa: support for a eurypterid–scorpion clade. *Lethaia* **32**, 72–74.

Briggs, D. E. G., Clarkson, E. N. K. and Aldridge, R. J. 1983. The conodont animal. *Lethaia* **16**, 1–14.

Briggs, D. E. G., Siveter, D. J. and Siveter, D. J. 1996. Soft-bodied fossils from a Silurian volcaniclastic deposit. *Nature* **382**, 248–250.

Cocks, L. R. M., Brunton, C. H. C., Rowell, A. J. and Rust, I. C. 1970. The first Lower Palaeozoic fauna proved from South Africa. *Quarterly Journal of the Geological Society of London* **125**, 583–603.

Cocks, L. R. M. and Fortey, R. A. 1986. New evidence on the South African Lower Palaeozoic: age and fossils reviewed. *Geological Magazine* **123**, 437–444.

Fortey, R. A. and Theron, J. N. 1995. A new Ordovician arthropod, *Soomaspis*, and the agnostid problem. *Palaeontology* **37**, 841–861.

Gabbott, S. E. 1998. Taphonomy of the Ordovician Soom Shale Lagerstätte: an example of soft tissue preservation in clay minerals. *Palaeontology* **41**, 631–667.

Gabbott, S. E. 1999. Orthoconic cephalopods and associated fauna from the late Ordovician Soom Shale Lagerstätte, South Africa. *Palaeontology* **42**, 123–148.

Gabbott, S. E., Aldridge, R. J. and Theron, J. N. 1995. A giant conodont with preserved muscle tissue from the Upper Ordovician of South Africa. *Nature* **374**, 800–803.

Gabbott, S. E., Aldridge, R. J. and Theron, J. N. 1998. Chitinozoan chains and cocoons from the Upper Ordovician Soom Shale Lagerstätte, South Africa: implications for affinity. *Journal of the Geological Society of London* **155**, 447–452.

Gabbott, S. E., Norry, M. J., Aldridge, R. J. and Theron, J. N. 2001. Preservation of fossils in clay minerals; a unique example from the Upper Ordovician Soom Shale, South Africa. *Proceedings of the Yorkshire Geological Society* **53**, 237–244.

Kovács-Endrödy, E. 1987. The earliest known vascular plant, or a possible ancestor of vascular plants, in the flora of the Lower Silurian Cedarberg Formation, Table Mountain Group, South Africa. *Annals of the Geological Survey of South Africa* **20**, 893–906.

Rayner, R. J. 1986. *Promissum pulchrum*: the unfulfilled promise? *South African Journal of Science* **82**, 106–107.

Selden, P. A. 1985. Eurypterid respiration. *Philosophical Transactions of the Royal Society of London*, Series B **309**, 219–226.

Theron, J. N., Rickards, R. B. and Aldridge, R. J. 1990. Bedding plane assemblages of *Promissum pulchrum*, a new giant Ashgill conodont from the Table Mountain Group, South Africa. *Palaeontology* **33**, 577–594.

THE HUNSRÜCK SLATE

Chapter Four

Background: the rise of the vertebrates and the age of fishes

Vertebrates (animals with backbones) are members of the phylum Chordata, which all share a notochord (a tough rod of collagen running down the length of the body), and V-shaped blocks of muscle called myotomes. It has long been recognized that the oldest chordate in the geological record is the tiny *Pikaia* from the Middle Cambrian Burgess Shale of British Columbia (Chapter 2) and thus that the chordates originated during the Cambrian Explosion.

However, the discovery in 1999 of small, jawless, agnathan fish from the Lower Cambrian of southern China (see Chapter 2, p.28), has pushed the origin of the vertebrates themselves back into the melting pot of the Cambrian Explosion. The jawless agnathans, which include the present-day lampreys and hagfish with their sucker-like mouths, are certainly a primitive group of vertebrates, but the ancestors of higher vertebrates are not to be found amongst them.

During the early Silurian a second group of fish appeared, the enigmatic acanthodians (inappropriately named the 'spiny sharks' because of their prominent caudal fin). Generally small and with numerous spines on the dorsal and ventral margins, they became abundant in Devonian freshwaters and persisted until the end of the early Permian. They are significant as they are the first jawed vertebrates in the fossil record.

However, from the Cambrian through the Silurian the fossil record of fish is very sparse. The major episodes of early fish evolution most probably took place at this time, but did so in freshwater habitats, and it is not until the Devonian Period that freshwater deposits (mainly on the Old Red Sandstone continent) become common in the geological record. The Devonian Period is known as 'the age of fishes' because by its close all five major groups of fish were locally abundant (see Benton, 2000 for details).

The third group to appear were the placoderms – heavily armoured fish with poorly developed jaws. They were almost confined to the Devonian and are now extinct. For a time, however, they were dominant and a huge variety of weird forms with massive head armour included giants such as *Dunkleosteus*, over 20 m (65 ft) long. Its jaws had no teeth, but the upper edges of the lower jaw were sharpened into blade-like plates.

Finally the two major extant groups of fish, the osteichthyans and the chondrichthyans, evolved and flourished at this time. Osteichthyans (the bony fish) appeared early in the Devonian and by the end of the Palaeozoic had almost sole possession of lakes and streams and had invaded the seas. They include two lineages: the sarcopterygians (lobe-finned fish), such as lungfish and coelacanths, which in Devonian times were more significant; and the actinopterygians (ray-finned fish), which today include almost all freshwater fish and the vast majority of marine forms.

Chondrichthyans (the cartilaginous fish) – the sharks, skates, rays, and the chimaeras (ratfish) – were the last of the five classes of fish to evolve. They are not known before the late Middle Devonian, suggesting that their lack of bone is not a primitive condition, but rather that evolution is towards bone reduction.

By the end of the Devonian the early dominance of seas and freshwater by the armoured fish had given way to the modern sharks and bony fish. Moreover, by the Middle Devonian Givetian Stage the sarcopterygians had given rise to a second major group of vertebrates, and the first to conquer land, when their lobe-fins, supported by a single basal bone and strong muscles, were modified into the tetrapod limbs of the first amphibians.

Several rich fossil deposits within Devonian rocks record these important chapters in vertebrate evolution, such as the late Devonian marine deposits of Gogo in Western Australia and the Old Red Sandstone lake deposit of Achanarras in Caithness, Scotland (p. 46). The Lower Devonian Hunsrück Slate of the Rhenish Massif in western Germany, however, not only includes a rich plant and invertebrate biota in addition to vertebrates, but is also a true Conservation Lagerstätte with soft tissues exquisitely preserved in many groups, most notably the echinoderms (especially the starfish) and arthropods. Moreover, their peculiar preservation by pyritization enables their examination by X-ray techniques which often reveal minute detail.

History of discovery and exploitation of the Hunsrück Slate

The Hunsrück Slate has been an important source of roof slates for several centuries in the Rheinisches Schiefergebirge (Slate Mountains) of the Rhine and Mosel valleys (**48**). Certainly the slate was used in Roman times as evidenced from numerous Roman sites in west Germany, but the earliest documented evidence of mining in this area comes from the fourteenth century (Bartels *et al.*, 1998). Extensive mining continued throughout the succeeding centuries, but the Industrial Revolution of the late eighteenth century saw a huge expansion in the production of roofing slates. Slate was exported along the Rhine and Mosel rivers, but a collapse in the industry during the German depression (1846–49) led to poverty and hardship in the mining settlements.

Economic revival and a new sense of nationalism after the Franco-Prussian war of 1870–71 led to renewed growth in the slate industry and extensive mines were developed by the bigger companies. In the early twentieth century deeper shafts were sunk, railways were employed and the use of technology generally increased. Production continued through the years of the Second World War into the 1960s, but competition from synthetic and cheaper imported slate has since led to a decline. In recent years only one mine (Eschenbach-Bocksberg Quarry) has been operating in the Bundenbach region and, since 1999, this has merely been preparing imported slate from Spain, Portugal, Argentina and China and has ceased mining local rock.

The mining of the Hunsrück Slate was vital to the discovery of its fossils. Although they are not uncommon, fossils are only readily recognised by handling large quantities of rock and many of the fine specimens on display in museums were originally found by the slate splitters. The first scientific paper on these fossils was by Roemer (1862) who described and illustrated asteroids and crinoids from the Bundenbach region. Distinguished German palaeontologists such as R. Opitz (1890–1940), F. Broili (1874–1946), R. Richter

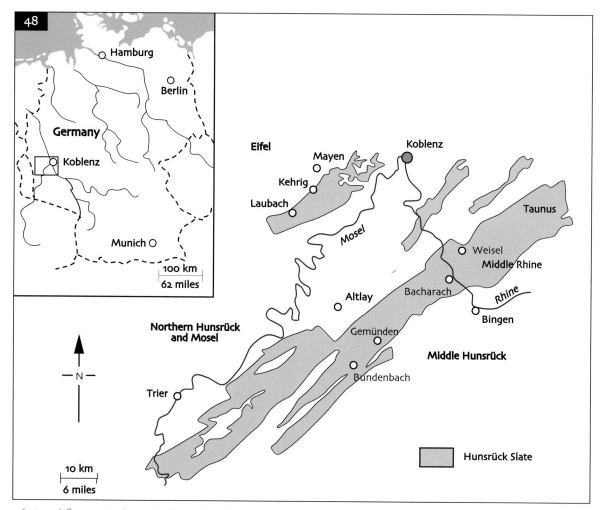

48 Locality map to show the Hunsrück Slate region in the Rheinisches Schiefergebirge of western Germany (after Bartels *et al.*, 1998).

(1881–1957) and W. M. Lehmann (1880–1959) made extensive studies of Hunsrück fossils from the 1920s to the 1950s, but Lehmann's death, coupled with the decline in the slate industry, led to a corresponding decline in research, especially as few new specimens were being discovered.

At the end of the 1960s Wilhelm Stürmer, a chemical physicist and radiologist at Siemens Corporation, combined these skills with his interest in palaeontology and developed new methods of examining the Hunsrück fossils using X-rays (Stürmer, 1970). His beautiful radiographs of unprepared slates, using soft X-rays (25–40 KV) and stereoscopic exposures combined with high-resolution films and image processing, show intricate detail of soft tissue not revealed by conventional techniques (**49, 50**). More recently Günther Brassel and Christoph Bartels have continued the work of Stürmer (Bartels and Brassel, 1990). Bartels *et al.* (1998) give an extensive bibliography for all of this research.

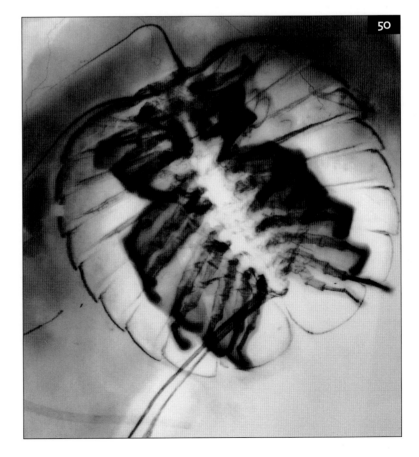

49 Radiograph of the starfish *Helianthaster rhenanus* (DBMB). Width approx. 150 mm (6 in).

50 Radiograph of the arthropod *Cheloniellon* (BKM). Width 120 mm (4.8 in).

Stratigraphic setting and taphonomy of the Hunsrück Slate

The Hunsrück Slate (Hunsrückschiefer) is a thick sequence of muddy marine sediments of Lower Devonian age which by low-grade metamorphism have been altered to slates. The sequence should strictly be regarded as a facies, rather than a stratigraphic unit, as the deposition of mud was diachronous, beginning earlier in the north-west and migrating south-eastwards. The precise age of the slate is therefore variable, but ranges from late Pragian to early Emsian and is therefore approximately 390 million years. Its outcrop in the Hunsrück hills (**51**, **52**) occurs in a belt approximately 150 km (90 miles) in length and covers an area of 400 square km (150 square miles).

Deposition of the mud occurred in a narrow north-east–south-west offshore marine basin lying between the recently uplifted Old Red Sandstone Continent to the north and the Mitteldeutsche Schwelle (see Chapter 10, p. 101) to the south. Immediately after its uplift, at the end of the late Silurian/early Devonian Caledonian Orogeny, large volumes of sand and muddy detritus were shed into rivers and transported south. Finer sediment was carried in suspension and deposited offshore in the Central Hunsrück Basin. The total thickness of the Hunsrück Slate has been estimated at 3,750 m (12,300 ft) (Dittmar, 1996), although the roofing-slate sequence in the area around Bundenbach and Gemünden is somewhat less than 1,000 m (3,300 ft) (**53**).

51 The Hunsrück hills near Fischbach in the Rheinisches Schiefergebirge of western Germany.

52 Underground mining of the Hunsrück Slate at Herrenberg Mine, Bundenbach.

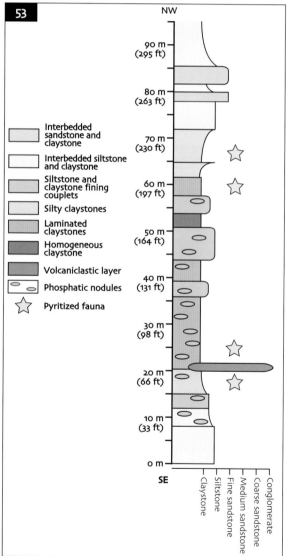

53 Generalized log of part of the Hunsrück Slate sequence in the Bundenbach area (after Sutcliffe et al., 1999).

During the subsequent Variscan Orogeny of the Carboniferous Period the muddy sediments were subjected to low-grade metamorphism, producing the characteristic slaty cleavage. On the limbs of the folds, such as in the Bundenbach–Gemünden area, the cleavage lies parallel to the bedding enabling the fossils to be revealed.

The ecological setting of the Hunsrück Slate biota has only recently been investigated in detail. The presence of photosynthesizing red algae and the well-developed eyes of certain fish and arthropods (Stürmer and Bergström, 1973; Briggs and Bartels, 2001), suggest that the community was living within the photic zone, i.e. less than 200 m (650 ft) below sea level (Bartels *et al.*, 1998). Average sedimentation rates have been estimated at only 2 mm (0.08 in) per year, but Brett and Seilacher (1991) suggested that rapid sedimentation occurred episodically, caused by tropical storms disturbing sediment in shallower water and sending sediment-laden turbidity currents down into deeper water. Communities living on the muddy sea floor were quickly overcome and buried, explaining the dominance of benthic organisms.

Earlier authors (e.g. Koenigswald, 1930) suggested that the currents transported communities from the area in which they were living into a hostile environment favourable for preservation, exactly as postulated for the Burgess Shale (Chapter 2). However, the burial of crinoids with rooting structures *in situ* and the preservation of syndepositional arthropod trails confirm that the Hunsrück community was living in the area in which it was fossilized, and that it was buried in life position (Sutcliffe *et al.*, 1999).

The bottom waters were clearly well oxygenated allowing the benthic community to become established, including a thriving infauna, as shown by diverse trace fossils. Preservation of soft tissue requires that it is not disturbed by burrowers after burial, so the sediment must have rapidly become anoxic and inhospitable, eliminating both benthos and scavengers. Burial events were, however, only of limited lateral extent, maybe confined to a few hundred square metres (several hundred square feet) (Bartels *et al.*, 1998, p. 50), and living communities survived adjacent to them.

The preservation of the Hunsrück Slate fossils is remarkable in that mineralized skeleton and un-mineralized soft tissue have both been preserved by the process of pyritization. Moreover, fragile skeletons (of echinoderms, for example) are often preserved as complete, articulated individuals. Soft tissue preservation, which includes arthropod limbs, eyes and intestines and the tentacles of cephalopods, is confined to four restricted horizons within the thick Hunsrück sequence in the Bundenbach Gemünden area only (**53**). Elsewhere the sequence yields the same taxa, but preserved as fragmented or disarticulated hard parts only. Conditions for rapid pyritization must only have existed for limited periods of time within a restricted area.

Pyritization of soft tissue is unusual in the fossil record; the only other Fossil-Lagerstätten to show such preservation are Beecher's Trilobite Bed (Ordovician) of New York State, and the Jurassic beds of La Voulte-sur-Rhône in France. Briggs *et al.* (1996) showed that the pyritization of soft tissue can only occur in exceptional circumstances of sediment chemistry when there is a low organic content, but a high concentration of dissolved iron. When a carcass is buried in such sediment, sulphate-reducing anaerobic bacteria break down its organic matter producing sulphide. The high concentration of iron in the sediment converts this to iron monosulphide. Finally, aerobic bacteria convert this by oxidation to pyrite. If the organic content of the sediment is too high the dissolved iron precipitates in the sediment and not in the carcass.

The pyrite that preserves the soft tissue of the Hunsrück Slate fossils thus grew as the tissue decayed, but unfortunately microstructure is not preserved as it is when the tissue is replaced by calcium phosphate (see Chapter 11, p. 114). Fossils are usually compressed, but pyritization does preserve some relief, which is not usually the case in argillaceous deposits such as the Burgess Shale (Chapter 2). It has been shown by Allison (1990) that, although the formation of pyrite initially involves reduction, it also requires oxidation at a later stage, and therefore occurs in the upper layers of the sediment near the aerobic–anaerobic interface.

Description of the Hunsrück Slate biota

Fish. As Bartels *et al.* (1998, p. 229) pointed out, the Hunsrück fossils provide an unrivalled picture of the diversity of fish in early Devonian seas. Four of the five main groups of fish are present with only the chondrichthyans, which had not yet evolved, unrepresented. Most of the fish present are agnathans and placoderms. Agnathans are represented by *Drepanaspis* (**54**), which is locally abundant and suggests temporary brackish conditions, and the much rarer *Pteraspis*. The flattened form of *Drepanaspis* suggests that it was nektobenthic and that it was easily caught in the

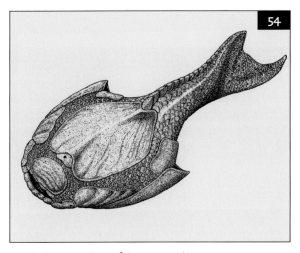

54 Reconstruction of *Drepanaspis*.

turbidity flows. Placoderms comprise several genera, such as *Gemuendina* (**55**, **56**) and *Lunaspis*, but are less common. The former was shaped like a modern ray and was probably also a bottom dweller – some individuals are up to 1 m (3 ft) long. Acanthodians are known from their fossilized spines up to 40 cm (16 in) long and are common in the Eifel region. Finally, a single specimen of a sarcopterygian (lobe-finned fish) represents the earliest record of a lungfish.

Echinoderms. 'Starfish' (including asteroids and ophiuroids) are perhaps the most familiar and most abundant Hunsrück fossils and are often found intact with the soft skin preserved between the arms. Asteroids (true starfish) comprise some 14 genera, most with five arms, but some (e.g. *Palaeosolaster*, **57**) had more than 20. *Helianthaster* (**49**, **58**) was one of the largest asteroids known, with 16 arms up to 20 cm (8 in) long. Ophiuroids ('brittle-stars'), also with 14 genera, are sometimes very common, such as the graceful *Furcaster* (**59**) and *Encrinaster* (**60**), with their long slender arms. Crinoids are also common (**61**), but

echinoids, blastoids, cystoids and holothuroids ('sea-cucumbers') are rare. Almost all of the 65 species of crinoids were sessile forms attached to the substrate and many of the Hunsrück specimens are preserved intact and articulated.

Annelids and arthropods. The polychaete worms ('bristle worms') of the Hunsrück Slate (e.g. *Bundenbachochaeta*, **62**) fill the gap between those of the Burgess Shale (Chapter 2) and Mazon Creek (Chapter 6); these entirely soft-bodied animals are rarely fossilized. Arthropods are spectacular with soft-part preservation of appendages and internal organs. The three major aquatic groups, trilobites, crustaceans and chelicerates, are all represented and there are also some enigmatic arthropods (such as *Mimetaster* with a star-shaped dorsal shield [**63**, **64**], and *Vachonisia* with a carapace like a large brachiopod; Stürmer and Bergström, 1976), which recall the more ancient body plans of some of the Burgess Shale arthropods (Chapter 2). They are important for they prove that successors of the archaic Burgess arthropods did survive at least until Devonian

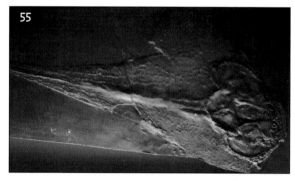

55 The placoderm fish *Gemuendina stuertzi* (SM). Length 220 mm (8.7 in).

56 Reconstruction of *Gemuendina*.

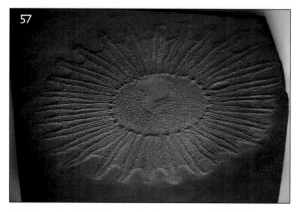

57 The starfish *Palaeosolaster gregoryi* (BKM). Width approx. 250 mm (10 in).

58 The starfish *Helianthaster rhenanus* (BM). Width approx. 150 mm (6 in).

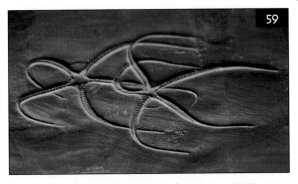

60 The brittle star *Encrinaster roemeri* (MM). Maximum width 120 mm (4.8 in).

59 The brittle star *Furcaster palaeozoicus* (BM). Length of arms approx. 75 mm (3 in).

61 The crinoid *Imitatocrinus gracilior* (BKM). Length of arms approx. 60 mm (2.4 in).

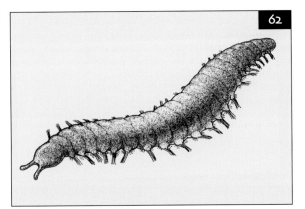

62 Reconstruction of *Bundenbachochaeta*.

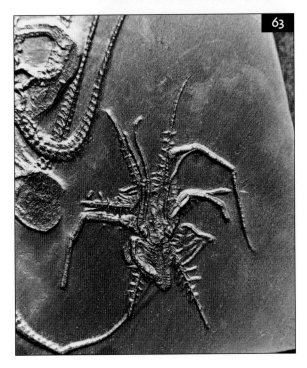

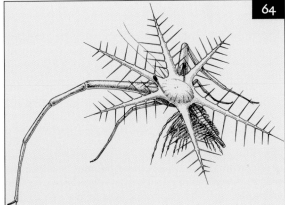

64 Reconstruction of *Mimetaster*.

63 The enigmatic arthropod *Mimetaster hexagonalis* (PC). Width of specimen (left to right) 46 mm (c. 2 in).

times (Briggs and Bartels, 2001). Trilobites are numerous, dominated by the phacopids (e.g. *Chotecops*, **65**), but crustaceans are much rarer, the most common being the bivalved malacostracan *Nahecaris* (**66, 67**). The chelicerates include rare xiphosurans, eurypterids and scorpions, as well as the only known fossil sea spiders (pycnogonids). The last group is represented in the Hunsrück fauna by the remarkable *Palaeoisopus* (**68, 69**) with a limb span of up to 40 cm (16 in).

Other invertebrates. None of the other invertebrate groups form dominant members of the Hunsrück fauna, but several groups are represented. Siliceous sponges (cf. *Protospongia*) are restricted to two genera, but cnidarians are more variable. They include chondrophorans ('by-the-wind-sailors'), solitary rugose corals (common Devonian types such as *Zaphrentis*, **70**), colonial tabulate corals (e.g. *Pleurodictyum* and *Aulopora*), scyphozoan conulariids and ctenophores ('comb jellies' or 'sea gooseberries'). Molluscs include gastropods, bivalves and cephalopods, the latter being an important element, mainly comprising orthoconic nautiloids and goniatites. Brachiopods (some with the soft pedicle preserved; see Südkamp, 1997) and bryozoans are also relatively common.

Plants. The calcareous alga *Receptaculites* is the only autochthonous marine plant. Fragments of terrestrial vascular plants also occur, washed out to sea from the coast, including members of the rhyniophytes (see Chapter 5).

Trace fossils. These include coprolites (from fish), epifaunal tracks (of arthropods, ophiuroids and fish), mobile infaunal traces (of bivalves, echinoderms and polychaetes), and constructed infaunal burrows (Sutcliffe *et al.*, 1999).

Palaeoecology of the Hunsrück Slate

The Hunsrück Slate, like the Burgess Shale (Chapter 2), represents a marine, benthic community living in, on, or just above the muddy seabed of an offshore basin situated at about 20°S. Bottom waters were oxygenated and subjected to currents and, as at Burgess, the presence of photosynthesizing algae suggests that the depth was certainly less than 200 m (650 ft).

No statistical analysis of the Hunsrück Slate biota has been attempted, but of the 400 species of macrofossils described, the majority were certainly benthic. A small percentage consisted of benthic infauna, living in the sediment itself, as shown by pyritized burrows such as *Chondrites*, and the infaunal traces of deposit feeders such as the polychaete worm *Bundenbachochaeta* together with some bivalves and echinoderms.

The majority of Hunsrück animals consist of benthic epifauna, living on the sediment surface, the sessile epifauna being dominated by meadows of crinoids, with sponges, corals, conulariids, brachiopods and bryozoans, plus most of the bivalves. These animals shared the seabed with calcareous receptaculitid algae. The vagrant epifauna, walking or crawling across the seabed, was dominated by starfish (asteroids and ophiuroids) and arthropods (trilobites, crustaceans, chelicerates and the archaic forms), but also included some gastropods.

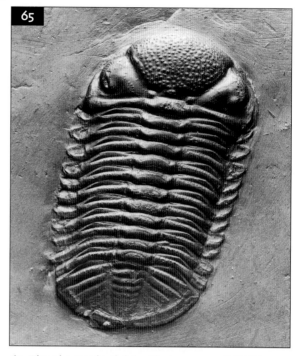

65 The phacopid trilobite *Chotecops* sp. (DBMB). Length 85 mm (3.4 in).

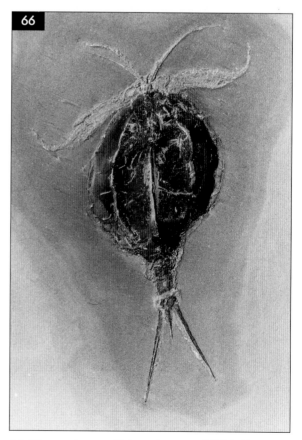

66 The crustacean *Nahecaris stuertzi* (DBMB). Total length 157 mm (6.2 in).

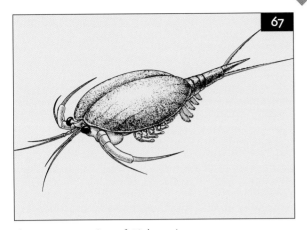

67 Reconstruction of *Nahecaris*.

68 The sea spider *Palaeoisopus problematicus* (BKM). Length of longest leg approx. 180 mm (7 in).

69 Reconstruction of *Palaeoisopus*.

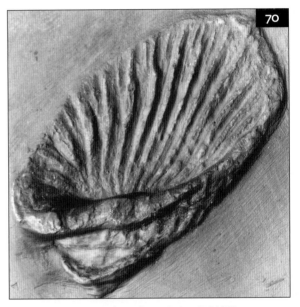

70 The rugose coral *Zaphrentis* sp. (DBMB). Width of calyx 48 mm (c. 2 in).

Animals living higher in the water column were generally able to escape the storm-induced mud flows, but nektobenthic animals (near-bottom swimmers) are represented by some of the agnathan and placoderm fish, such as *Drepanaspis* and *Gemuendina*, with their flattened, ray-like bodies. Planktonic floaters include the cnidarian chondrophorans and ctenophores, while the active nektonic swimmers include orthocone nautiloids, goniatites, acanthodian fish and the arthrodire placoderms up to 2 m (6 ft) long.

Trophic analysis identifies the full range of feeding types including: filter feeders, dominated by the crinoids and sponges; deposit feeders, including the gastropods, polychaetes, some arthropods (such as the enigmatic *Mimetaster* and *Vachonisia*), and possibly the starfish (their large mouths suggests that they were deposit feeders unlike modern predators: Bartels *et al.*, 1998, p. 43). Other arthropods were either scavengers, such as the phyllocarid crustacean *Nahecaris*, with its robust mandibles, or predators such as the giant sea spider, *Palaeoisopus*, which was armed with large chelicerae, or pincers, and probably preyed on crinoid meadows (Bergström *et al.*, 1980). The largest predators were undoubtedly the orthoconic nautiloids, plus the shark-like acanthodians and the arthrodire placoderms, which possibly preyed on orthocones. All of the cnidarians would have captured small organisms with their tentacles.

Comparison of the Hunsrück Slate with other Devonian fish beds
Gogo Formation, Western Australia

This locality in the Kimberley district of north-west Australia has yielded an exceptional fauna of fossil fish from a late Devonian marine (reef) setting. First discovered in the 1940s, the fossils are preserved in limestone concretions which formed during early diagenesis, thus preventing compaction and preserving the fish uncrushed in three dimensions (compare with

Santana Formation, Chapter 11). Careful preparation using acetic acid can free the fish entirely from their matrix. Preservation is spectacular and the fauna includes abundant armoured placoderms, including a new group, the camuropiscids, shark-like predators with a torpedo-shaped skull and tooth plates designed for crushing. Several new ray-finned fish and lungfish were also discovered during the Gogo Expedition of 1986 (see Long, 1988). Crustaceans are also common in the concretions and were probably the prey of the placoderms. As yet no acanthodians or cartilaginous fish have been discovered.

Achanarras Fish Bed, Caithness, Scotland

Fossil fish from the Devonian Old Red Sandstone of Scotland have been well known since Agassiz's classic work, *Recherches sur les Poissons Fossiles*, of the 1830s. They were deposited in the Orcadian Lake, a huge subtropical lake on the southern margins of the Old Red Sandstone Continent, which covered much of Caithness, the Moray Firth, Orkney and Shetland during the Middle Devonian. The fish lived in the shallow water of the lake margins where waters were warm and well oxygenated. On death their carcasses drifted to the centre of the lake and sank into deeper, colder, anoxic waters beneath a thermocline, where they were preserved in laminated muds on the lake floor, a situation not unlike that of the Crato Formation (Chapter 11).

Mass mortality events, possibly caused by deoxygenation of the water due to algal blooms, led to high concentrations of fish carcasses on the lake floor, which were preserved in good condition due to the lack of scavengers on the lake bed. The fauna includes agnathans, placoderms, acanthodians and bony fish including ray-finned and lungfish. Small arthropods may have been a food source for the smaller fish which themselves fell prey to the large carnivorous placoderms and lungfish (Trewin, 1985, 1986).

Further Reading

Allison, P. A. 1990. Pyrite. 253–255. *In* Briggs, D. E. G. and Crowther, P. R. (eds.). *Palaeobiology: a synthesis.* Blackwell Scientific Publications, Oxford, xiii + 583 pp.

Bartels, C. and Brassel, G. 1990. *Fossilien im Hunsrück-schiefer. Dokumente des Meereslebens im Devon.* Museum Idar-Oberstein Series 7, Idar Oberstein, 232 pp.

Bartels, C., Briggs, D. E. G. and Brassel, G. 1998. *The fossils of the Hunsrück Slate.* Cambridge University Press, Cambridge, xiv + 309 pp.

Benton, M. J. 2000. *Vertebrate palaeontology.* Blackwell Science, Oxford, xii + 452 pp.

Bergström, J., Stürmer, W. and Winter, G. 1980. *Palaeoisopus, Palaeopantopus* and *Palaeothea,* pycnogonid arthropods from the Lower Devonian Hunsrück Slate, West Germany. *Paläontologische Zeitschrift* **54**, 7–54.

Brett, C. E. and Seilacher, A. 1991. Fossil Lagerstätten: a taphonomic consequence of event sedimentation. 283–297. *In* Einsele, G., Ricken, W. and Seilacher, A. (eds.). *Cycles and events in stratigraphy.* Springer-Verlag, Berlin. xix + 955pp.

Briggs, D. E. G and Bartels, C. 2001. New arthropods from the Lower Devonian Hunsrück Slate (Lower Emsian, Rhenish Massif, western Germany). *Palaeontology* **44**, 275–303.

Briggs, D. E. G., Raiswell, R., Bottrell, S. H., Hatfield, D. and Bartels, C. 1996. Controls on the pyritization of exceptionally preserved fossils: an analysis of the Lower Devonian Hunsrück Slate of Germany. *American Journal of Science* **296**, 633–663.

Dittmar, U. 1996. Profilbilanzierung und Verformungsanalyse im südwestlichen Rheinischen Schiefergebirge. Zur Konfiguration, Deformation und Entwicklungsgeschichte eines passiven variskischen Kontinenentalrandes. *Beringia. Würzburger Geowissenschaftliche Mitteilungen* **17**, 346 pp.

Koenigswald, R. von. 1930. Die Fauna des Bundenbacher Schiefers in ihren Beziehungen zum Sediment. *Zentralblatt für Mineralogie, Geologie und Paläontologie* **B**, 241–247.

Long, J. A. 1988. The extraordinary fishes of Gogo. *New Scientist* **120** (1639), 40–44.

Roemer, C. F. 1862. Asteriden und Crinoiden von Bundenbach. *Verhandlungen der Naturhistorischen Vereins der Preussischens Rheinland und Westfalens. Bonn* **20**, 109.

Stürmer, W. 1970. Soft parts of cephalopods and trilobites: some surprising results of x-ray examination of Devonian slates. *Science* **170**, 1300–1302.

Stürmer, W. and Bergström, J. 1973. New discoveries on trilobites by x-rays. *Paläontologische Zeitschrift* **47**, 104–141.

Stürmer, W. and Bergström, J. 1976. The arthropods *Mimetaster* and *Vachonisia* from the Devonian Hunsrück Shale. *Paläontologische Zeitschrift* **50**, 78–111.

Südkamp, W. H. 1997. Discovery of soft parts of a fossil brachiopod in the 'Hunsrückschiefer' (Lower Devonian, Germany). *Paläontologische Zeitschrift* **71**, 91–95.

Sutcliffe, O. 1997. The sedimentology and ichnofauna of the Lower Devonian Hunsrück Slate, Germany: taphonomy and palaeobiological significance. Unpublished Ph.D. thesis, University of Bristol.

Sutcliffe, O. E., Briggs, D. E. G. and Bartels, C. 1999. Ichnological evidence for the environmental setting of the Fossil-Lagerstätten in the Devonian Hunsrück Slate, Germany. *Geology* **27**, 275–278.

Trewin, N. H. 1985. Mass mortalities of Devonian fish – the Achanarras Fish Bed, Caithness. *Geology Today* **2**, 45–49.

Trewin, N. H. 1986. Palaeoecology and sedimentology of the Achanarras fish bed of the Middle Old Red Sandstone, Scotland. *Transactions of the Royal Society of Edinburgh: Earth Sciences* **77**, 21–46.

The Rhynie Chert

Background: colonization of the land

The consequences of plants and animals leaving the marine realm and colonizing the surface of the land were far-reaching, not least for the evolution of the human race, which belongs to the Earth's terrestrial biota. Only very few of the metazoan phyla which emerged from the great Cambrian Explosion produced terrestrial forms, but these have been very successful in terms of diversity. The Arthropoda includes some terrestrial crustaceans (e.g. woodlice), but of much greater importance are the terrestrial chelicerates (spiders, scorpions, mites and their allies) and the insects (which make up 70% of all animals alive today). From fish arose the tetrapods – amphibians, reptiles, birds and mammals. Molluscs, too, in the form of slugs and snails, have been remarkably successful on land, as any gardener will testify. In order to live successfully on land, plants developed features such as stiff trunks and reproductive devices which gave rise to the familiar trees and flowers we see on land today. Terrestrialization was thus a major episode in the evolution of life on Earth. It was neither an instantaneous event nor restricted to a particular geological period; indeed, some animals (such as crabs) may be considered to be terrestrializing today. Nevertheless, when the physical conditions on the land surface became sufficiently favourable to life in the Silurian Period, the invasion began in earnest.

Among the physical barriers to be overcome when a plant or animal adjusts to life on land, having previously lived in the sea, is that of water supply. Water is necessary for all biological processes but its supply is variable on land, compared to the sea. Plants and animals adopt four main strategies to cope with under- (or over-) supply of water. Some, like microbes, ostracods and algae, live in permanent water on land (between soil particles and in ponds); they are effectively aquatic. Others, like amphibians, millipedes, slugs and woodlice, live in moist habitats and only venture out into dry air for short periods. Some plants and animals are poikilohydric, that is they can tolerate desiccation and rehydrate when necessary; examples are bryophytes (mosses and liverworts), and 'resting' stages such as plant spores, fairy shrimp eggs and tardigrade ('water-bear') tuns. This group was probably the first onto land. Finally, the most successful of all are the homoiohydric forms, which maintain permanent internal hydration mainly by having a waterproof cuticle or skin. These are the familiar land plants (tracheophytes), tetrapods and arthropods mentioned previously.

Another physiological barrier to be overcome when crossing the threshold from water to land is breathing or, more specifically, exchange of the gases oxygen (O_2) and carbon dioxide (CO_2). Both plants and animals exchange O_2 and CO_2 with the outside across a semi-permeable membrane but, because the O_2 and CO_2 molecules are both larger than the water molecule (H_2O), this membrane will leak, resulting in a loss of water. To overcome this problem, land plants have small holes (stomata) in their waterproof cuticle, which can be closed to prevent excessive water loss. Animals, which might have used external gills when living in water, have lungs or tracheal systems (in insects, for example) for breathing air enclosed within the body, and connected to the air by small holes (called spiracles in insects).

Other adaptations which evolved during terrestrialization include: the development of strong legs and better balance to compensate for the loss of buoyancy; sense organs which would operate in a medium which had different optical and acoustic properties (sound is used more for communication on land); more careful ionic balance, which is linked with the reduced availability of water on land; and the development of direct copulation in mating (in water gametes can simply be emptied close to the opposite sex without contact). In spite of all the potential problems, organisms swarmed onto the land. It was, after all, an unexploited ecological niche and there was, at least to begin with, some respite from predation in the sea.

History of discovery of the Rhynie Chert

The Rhynie Chert was discovered in 1912 by Dr William Mackie, who studied at the University of Aberdeen in Scotland and later became a physician and keen amateur geologist based in Elgin. In a field outside the Aberdeenshire village of Rhynie (**71**) he found lumps of chert, a siliceous rock produced by hot springs, containing obvious plant axes (stems) and rhizomes (underground stems) (**72**, **73**). Thin sections of the rock reveal exquisite preservation of cells, including the water-conducting vessels typical of vascular land plants (**74**). In October 1912 a trench was dug by Mr D. Tait, a fossil collector to the Geological Survey and, as a result of the material discovered, between 1917 and 1921 a series of five papers was published by Kidston and Lang in which they described the plants *Rhynia*, *Aglaophyton*, *Horneophyton* and *Asteroxylon*. A period of some 30 years went by without any major contributions, but in the late 1950s interest was rekindled through the work of Dr A. G. Lyon of Cardiff University in Wales, who described spores fossilized in the process of germination. Further trenches were dug in the 1960s and 1970s. Dr Lyon later bought the Rhynie Chert site and donated it to Scottish Natural Heritage in 1982. There is little evidence of these trenches to be seen at Rhynie today; just a grassy field and grazing livestock. Now, in addition to the vascular plants, algae, fungi and lichens are known from the site, the new material having been discovered by a team of palaeobotanists at the University of Münster in Germany, led by the late Professor Remy.

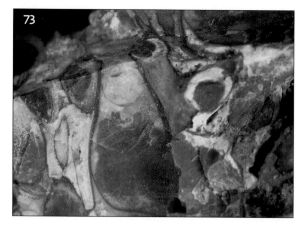

71 The Rhynie Chert, Aberdeenshire, Scotland, lies beneath this field. The hills behind are formed of older rocks beyond the inlier.

72 A piece of Rhynie Chert, about 120 mm (5 in) long.

73 Closer view of the piece of Rhynie Chert shown in Figure **72**. Note the tubular structures, about the size of macaroni, which are stems of vascular plants.

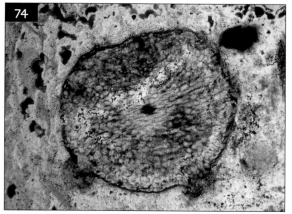

74 Thin section of Rhynie Chert showing detailed preservation of cellular structure in vascular plant stems. Stem is about 2 mm (0.08 in) across.

Shortly after the discovery of the plants, in the 1920s animal remains were described from the chert. Mites and other arachnids (trigonotarbids) were reported by Hirst (1923), fairy shrimps by Scourfield (1926) and springtails by Hirst and Maulik (1926). With few exceptions, the accuracy of the descriptions and the detailed drawings of arthropods in these papers has been corroborated by later workers. In 1961 Claridge and Lyon described book-lungs in the trigonotarbid arachnids, thus proving that they were, indeed, land animals. More recently, centipedes have been found in the chert, as well as an arthropod with gut contents which tell us that it was a detritus feeder (Anderson and Trewin, 2002).

While early studies concentrated on the remarkably well-preserved early land plants and animals – the site held the record for the earliest land animals for some 70 years – little work was done on the geological aspects until recently. Interest in the area was renewed in 1988 when Rice and Trewin from the University of Aberdeen demonstrated that the silicified rocks in the area are enriched with gold and arsenic, thus confirming their hot-spring origin. The ensuing mineral exploration revealed much about the subsurface geology but little exploitable gold. The discovery by the Aberdeen group in the 1990s of a new fossiliferous chert, the Windyfield Chert, some 700 m (c. 0.5 mile) from the original locality, resulted in recognition of part of a geyser vent rim. Work is continuing, in part by comparison with modern silicious hot-spring systems such as those at Yellowstone National Park, Wyoming (**75**, **76**).

75 Castle Geyser, Yellowstone National Park, Wyoming. The white deposit around the geyser (hot spring) is sinter; the different colours result from different species of algae and cyanobacteria, which each live in different water temperatures.

76 The plant *Triglochin*, which tolerates life in warm water from hot springs and, though only distantly related botanically, resembles the kind of vegetation which grew around the Rhynie hot spring. Fountain Paint Pots, Yellowstone National Park, Wyoming. The plant grows up to 30 cm (12 in) tall.

Stratigraphic setting and taphonomy of the Rhynie Chert

The host rocks of the Rhynie chert are Early Devonian (Pragian; c. 396 Ma) in age, and are included in a sequence of sedimentary rocks deposited by streams, rivers and lakes during the Devonian Period when Scotland, much of northern Europe, Greenland and North America were joined together in a large continent called Laurasia, located between 0° and 30° south of the equator. At Rhynie, the Devonian rocks are surrounded by older Dalradian metamorphic and Ordovician plutonic igneous rocks (**77**, **78**). The Rhynie sediments were deposited in a relatively narrow, north-east–south-west trending basin within these older rocks. The basin is a half-graben; the western edge is marked by a fault which was active at the time of deposition, and at the eastern edge the sedimentary rocks lie unconformably on basement rocks. The palaeoenvironment envisaged for deposition of the chert is one of rivers and lakes depositing a complex of cross-bedded sands in the high-flow regions and muds on floodplains and in shallow ephemeral lakes. Hydrothermal activity centred on the fault system altered the sub-surface rocks in the vicinity of the fault and deposited sinter around hot springs and geysers at the surface in the Rhynie area.

Later Earth movements have caused strata within the basin to dip towards the north-west, while the chert-bearing rocks near Rhynie village are folded into a syncline which plunges to the north-east.

The preservation of the biota is variable, and depends mainly on two things: the condition of the organisms at the time of fossilization (the amount of decay) and the degree and timing of silica replacement. There is a whole array of preservation types, ranging from the exquisitely preserved three-dimensional internal anatomy of those organisms that were completely silicified at or soon after death, to organisms preserved as compressed, unidentifiable, coalified strips silicified after decay and burial. In a few cases where the whole bodies (cf. moulted skins) of arthropods have been found, gut contents have been preserved, and even delicate features such as book-lungs are perfectly silicified. Some arthropod fossils clearly represent moults: Figure **79** shows a cross-section of a trigonotarbid opisthosoma (abdomen) containing a leg! The leg moult must have become lodged in the shed skin of the opisthosoma during moulting. Orientation of the plants varies from upright stems with terminal sporangia (spore-cases) and horizontal rhizomes (therefore presumably in growth position) to collapsed, decayed, prostrate straws which form a layer of litter. Some plant sporangia have been found to

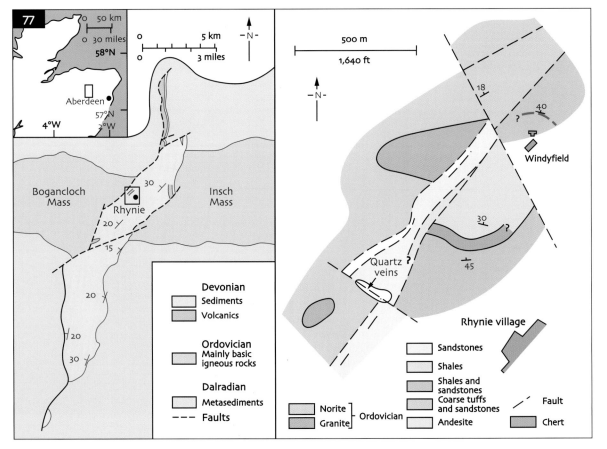

77 Locality map of the Rhynie Chert, Aberdeenshire, Scotland (after Rice *et al.*, 2002).

contain arthropod remains (**80**), but because these sporangia are among the horizontal plant debris, it is thought most likely that the animals entered empty (dehisced) sporangia when they were lying on the ground, as a sheltered place to moult, rather than climbing the plants to feed on the spores.

In hot-spring environments silica is precipitated as sinter: an amorphous, hydrated opaline form called Opal-A. When the hydrothermal solution, super-saturated with respect to Opal-A, erupts at the surface as a geyser or hot spring it cools and Opal-A is precipitated. As well as the drop in temperature and evaporation of the hydrothermal fluids, other factors may affect precipitation, such as pH, and the presence of dissolved minerals, organic matter and living organisms such as cyanobacteria. Silicification of plant material is a permeation and void-filling process called permineralization, in contrast to the direct replacement of cell walls (petrifaction) in which the organic structure acts as a template for silica deposition. Silicic acid (the common soluble form of silica) polymerizes with the loss of water, and opaline silica forms.

78 Stratigraphy of the Rhynie Chert, Aberdeenshire, Scotland (after Rice *et al.*, 2002).

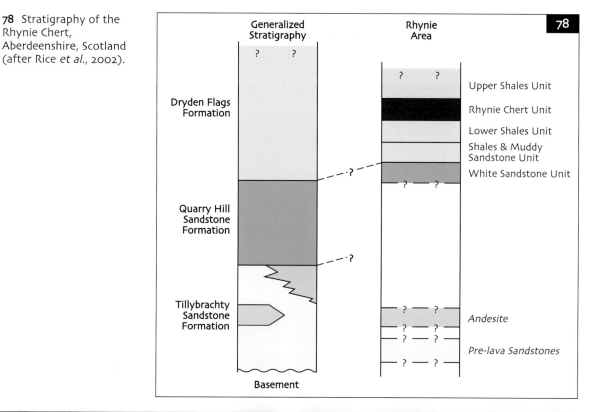

79 Cross section of abdomen of a trigonotarbid arachnid containing a trigonotarbid leg, thus proving that this specimen is a moulted skin. Abdomen about 1.5 mm (0.06 in) across.

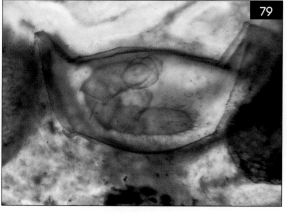

80 Trigonotarbid remains (including pedipalp, left, and chelicera) inside the sporangium of a vascular plant. Note the thickened sporangium wall (c. 0.5 mm [0.02 in] thick).

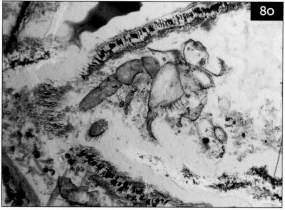

Nucleation of amorphous silica on wood and plant material involves hydrogen bonding between the hydroxyl groups in the silicic acid, and cellulose and lignin within the organic tissue. Silica precipitation within the cells and openings between cell walls preserves the plant's histology. The initial phase of silicification of organic matter may occur in a matter of days, but for complete silicification a regular, high influx of silica-bearing solutions is required. In the Rhynie hydrothermal area, regular outflow of solutions from hot springs and geysers produced the rapid and pervasive silicification before any significant cellular decay, and has led to the exquisite preservation of the plants in the Rhynie Chert. Sinter is a light, porous substance which bears little resemblance to chert rock. However, during burial and over time the amorphous silica phase Opal-A becomes unstable and gradually changes to the more stable crystalline form of silica: quartz. As the sinter is being transformed from Opal-A to quartz, percolating silica-bearing fluids may precipitate yet more crystals in voids and fractures in the rock so that the resulting chert retains almost none of its original porosity.

Description of the Rhynie Chert biota

In many layers of the chert, the plants are preserved so well that details of their cellular structure can be studied (74). These early land plants are relatively simple in their level of organization and include seven tracheophytes (higher land plants) and their relatives, two nematophytes (an enigmatic group), and a number of other types including fungi, algae and the oldest known fossil lichen.

It can be demonstrated that at least seven of the plants are true land plants by the following features: cuticles, stomata, intercellular air spaces (for gas diffusion), vascular strands with lignin (for water conduction and support), sporangia with a well-developed dehiscence, and spores. Five of these plants are true vascular plants or tracheophytes which have tracheids in the water-conducting cells. All have a simple construction and few exceeded 200 mm (8 in) in height. They resemble the living primitive plant *Psilotum* (**81**, **82**). They show two generations: the asexual, sporangium-bearing sporophyte, and the sexual (haploid) gametophyte. Gametophytes of some of these plants are known.

Aglaophyton major (**83**) has creeping rhizomes and smooth, naked, upright axes up to 6 mm (0.24 in) in diameter. The rhizomes show bulges bearing tufts of rhizoids for taking up water and nutrients. Branching is dichotomous and fertile axes terminate in pairs of cigar-shaped sporangia. The male gametophyte is known as *Lyonophyton rhyniensis*; it was much smaller and the upright axis ended in a cup-like structure that bore the antheridia (male organs). The systematic position of *Aglaophyton* is uncertain because it shows a mixture of features of different groups of plants. It has many characteristics of the rhyniophytes, a group of primitive plants known only from fossils, which have simply branched, naked stems (*Rhynia* is an example). The vascular cells (xylem) do not show thickenings and are more reminiscent of the hydroids of bryophytes (mosses and liverworts). The xylem of *Aglaophyton* does not possess true tracheids, which suggests it is not a

81 The living, primitive, leafless vascular plant *Psilotum nudum*, with lateral, subspherical sporangia, which shows the grade of organization of the Rhynie Chert vascular plants. Waimangu Valley, Rotorua, New Zealand. The plant is about 100 mm (4 in) tall.

82 Top view of *Psilotum nudum*, showing dichotomous branching pattern

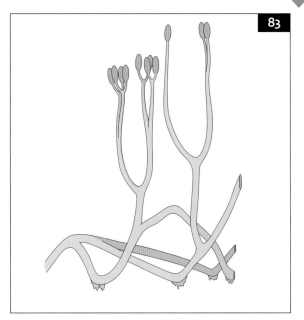

83 Reconstruction of *Aglaophyton*. Maximum height about 200 mm (8 in).

84 Reconstruction of *Asteroxylon*. Maximum height about 400 mm (16 in).

85 Reconstruction of *Horneophyton*. Maximum height about 200 mm (8 in).

true tracheophyte. *Aglaophyton* apparently grew mainly on dry, litter-covered substrates, on its own or with other plants, though it probably required wet conditions for germination.

Asteroxylon mackiei (**84**) is one of the more advanced and complex Rhynie plants. It had an extensive system of branching rhizomes; the upright axes grew to about 400 mm (16 in) in height with a maximum diameter of 12 mm (0.5 in) and with dichotomous branching. The aerial axes possessed scale-like enations ('leaves'). On the fertile axes, each kidney-shaped sporangium was attached by a stalk between an enation and the axis. The water-conducting strand of *Asteroxylon* is characteristically star-shaped in cross-section (hence its Latin generic name); smaller strands radiate from each star point to meet the bases of the enations. *Asteroxylon* belongs to the lycophytes, a group which includes modern club-mosses. It grew mainly in organic-rich soils as part of a diverse community together with other plants, and could probably tolerate quite dry habitats.

Horneophyton lignieri (**85**) has smooth, naked, upright axes and a bulbous rhizome bearing tufts of rhizoids. The aerial axes grew up to 200 mm (8 in) in height, had a maximum diameter of 2 mm (0.08 in), and showed repeated dichotomous branching. Fertile axes terminated in branched, tubular sporangia with a central columella. The female gametophyte stage of *Horneophyton*, *Langiophyton mackiei*, was much smaller, the upright axis ending in a cup-like structure that bore the archegonia (female organs). The vascular cells in *Horneophyton* possessed thickenings, which suggest it was a tracheophyte. However, the presence of a columella in the sporangium shows similarities with bryophytes. *Horneophyton* seemed to prefer sandy, organic-rich substrates, often on its own, and flourished in damp to wet conditions.

Nothia aphylla (**86**) had an extensive, branching network of rhizomes with ventral ridges bearing rhizoid tufts. Branches turned upright to form the aerial axes which were naked but had an irregular surface. Repeated dichotomous branching gave the plant a dense growth form. Fertile axes bore lateral, kidney-shaped, stalked sporangia. The male gametophyte, *Kidstonophyton discoides*, was much smaller, with upright axes bearing cup-like structures with tubular outgrowths bearing the antheridia. The vascular cells did not have thickenings and were similar to those seen in some modern bryophytes, the sporangia are comparable to those of a group of primitive fossil plants called zosterophylls, and the simply branched naked axes resemble those of rhyniophytes, so its systematic position is uncertain. *Nothia* grew in sandy soils and plant litter, on its own or with other plants.

Rhynia gwynne-vaughanii (**87**) was one of the commonest plants in the Rhynie ecosystem. Like *Aglaophyton*, it had creeping, branched rhizomes and smooth, naked upright axes. It grew to a height of about 200 mm (8 in), with dichotomously branching axes up to 3 mm (0.1 in) in diameter. *Rhynia* possessed curious hemispherical projections on its axes, and those on the rhizomes bore tufts of rhizoids. Fertile axes had terminal cigar-shaped sporangia. *Rhynia* is the typical member of an extinct group of primitive plants called rhyniophytes, which are characterized by their simple branching and naked stems. *Rhynia* commonly grew in thickets, typically on its own, and was also an early colonizer of well-drained sinter and sandy substrates. It is also found with other plants in a wide range of habitats.

Trichopherophyton teuchansii is a rare plant in the Rhynie ecosystem. Its height is unknown but the dichotomously branched, aerial axes had a maximum diameter of 2.5 mm (0.1 in). The subterranean rhizomes were smooth with small, blunt structures which probably acted as rhizoids. The aerial axes bore spiny projections. Fertile axes bore lateral, stalked, kidney-shaped sporangia, also with spiny projections. The vascular cells in *Trichopherophyton* possessed thickenings which, together with the shape and position of the sporangia, suggests it belongs to the zosterophylls. *Trichopherophyton* was a late colonizer of organic-rich substrates, always growing with other plants in a diverse flora.

Ventarura lyonii is a recently discovered higher land plant from the Windyfield Chert. Its height is uncertain but was at least 120 mm (5 in). The repeatedly dichotomously branched aerial axes had a maximum diameter of 7.2 mm (0.28 in). The subterranean rhizomes were smooth with small blunt-tipped structures acting as rhizoids. The aerial axes bore peg-like projections. The axes internally show a lignified middle layer to the cortex called the sclerenchyma. Fertile axes bore lateral, stalked, kidney-shaped sporangia. The vascular cells possessed thickenings which, together with the type of sporangia, suggest *Ventarura* belongs to the zosterophylls. The palaeoecology of *Ventarura* is not generally known, but it probably grew in patches near freshwater ponds and probably in sandy and organic-rich substrates.

Other plants. Nematophytes are an extinct group of plants consisting of an inner lattice of spirally coiled tubes which may be smooth or spirally thickened. The tubes are closely packed near, and may be orientated perpendicular to, the edge of the plexus where there is an outer cuticular envelope. Their internal structure shows similarities to those of certain algae, and the spirally thickened tubes resemble the tracheids of vascular plants. *Nematophyton taiti* and *Nematoplexus rhyniensis* are known from Rhynie but both are poorly preserved. The common habitat of nematophytes is not known, but they may have been semi-aquatic plants with emergent fronds.

86 Reconstruction of *Nothia*. Maximum height about 200 mm (8 in).

87 Reconstruction of *Rhynia*. Maximum height about 200 mm (8 in).

Cyanophytes or cyanobacteria are simple, photosynthetic bacteria. They may be unicellular or form filamentous chains of cells. They are prokaryotes, so the cells do not contain nuclei. A number of probable cyanobacteria are found in the Rhynie Chert, some forming distinct stromatolitic laminae, possibly having grown up from cyanobacterial mats on sinter surfaces. Other types are found in more aquatic parts of the chert, and still others within decaying plants. Some of the Rhynie cyanobacteria may have played a significant role in fixing atmospheric nitrogen in the soil.

Chlorophytes or green algae are photosynthetic eukaryotes (their cells contain nuclei). They may be unicellular, form filamentous chains of cells, or more complex structures as in the stoneworts (see below). A number of filamentous and unicellular green algae are known from the Rhynie Chert, particularly in chert beds deposited in aquatic environments, although commonly their poor preservation makes identification difficult.

Charophytes (stoneworts) are large, structurally complex, green algae comprised of a series of multicellular nodes and long single cells or internodes. Branching occurs at the nodes and may be repeated. Charophytes occur in fresh to brackish water. One probable charophyte has been described from the Rhynie Chert, *Palaeonitella cranii*, but its reproductive structures have not been discovered and therefore its identity cannot be proven. *Palaeonitella* is commonly found in aquatic parts of the chert along with the crustacean *Lepidocaris*.

Fungi are multicellular, non-photosynthetic eukaryotes. They feed saprophytically (on dead organic matter), parasitically (on live organisms), symbiotically with green plants (endotrophic mycorrhizae), or with an alga or cyanobacterium in a lichen. Numerous fungi are recorded from the Rhynie Chert, including the earliest good examples of endotrophic mycorrhizae in plant tissue.

Lichens are non-vascular organisms formed by the symbiotic relationship between a fungus and an alga or a cyanobacterium. The lichen thallus comprises distinct layers of fungal hyphae (the mycobiont) and the alga/cyanobacterium (the photobiont). The oldest known lichen is recorded from the Rhynie Chert: *Winfrenatia reticulata*. *Winfrenatia* most likely colonized hard substrates such as degrading sinter surfaces. It may have weathered rock surfaces and thus contributed to soil formation.

The Rhynie faunal list consists of crustaceans, trigonotarbid arachnids, mites, collembolans, euthycarcinoids and myriapods.

Crustaceans. The commonest arthropod in the Rhynie Chert is *Lepidocaris rhyniensis* (**88**). It was first described by Scourfield (1926), who erected a new crustacean order, Lipostraca, for the new animal. He later described young stages of the animal (Scourfield, 1940). *Lepidocaris* is a tiny, multi-segmented form with 11 pairs of phyllopods (leaf-like limbs), long, branched antennae and a pair of caudal appendages. It was aquatic, living on detritus in ephemeral pools in the hot-spring environment like fairy shrimps do today.

Chelicerates. This group of arthropods is characterized by the presence of a pair of chelicerae (small claws or fangs) in front of the mouth, and no antennae. Trigonotarbids (**79, 80, 89–91**) are extinct arachnids similar in appearance to spiders but lacking the definitive features of poison glands and silk-producing

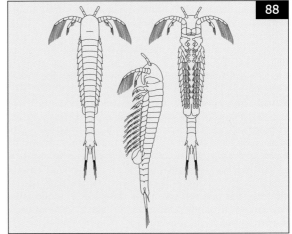

88 Reconstruction of *Lepidocaris*. Length about 4 mm (0.16 in) (after Scourfield, 1940).

89 The trigonotarbid arachnid *Palaeocharinus* in approximately sagittal section (NHM). Whole animal was about 2 mm (0.08 in) long.

90 Leg of *Palaeocharinus*, about 0.5 mm (0.02 in) long. Note the claw arrangement.

91 Reconstruction of *Palaeocharinus*. Body length about 3 mm (0.1 in).

organs, and show features which are primitive to spiders such as segmentation in the opisthosoma (abdomen). Their remains are common in the Rhynie Chert, and were first described as *Palaeocharinus rhyniensis* in the 1920s by Hirst and by Hirst and Maulik. The discovery of well preserved book-lungs (internal air-breathing organs connected to the outside by small spiracles) in Rhynie trigonotarbids by Claridge and Lyon (1961) removed any possible doubt that these were truly terrestrial air-breathers. *Palaeocteniza crassipes* was described as a spider by Hirst (1923), but restudy by Selden *et al.* (1991) concluded that this was erroneous and that *Palaeocteniza* is probably a moult of a juvenile trigonotarbid. Trigonotarbids were carnivores, as are all arachnids except some parasitic mites, presumably feeding on any animals they could catch. Like spiders, having caught their prey they would have poured digestive fluid into it through holes made by cheliceral fangs and then sucked out the liquified flesh.

The world's oldest known mites (Acari) occur in the Rhynie Chert (**92**). Like spiders and trigonotarbids, mites are arachnids, but they are very small and their prosoma and opisthosoma are not clearly defined. Hirst thought the Rhynie specimens all belonged to the same species, which he named *Protacarus crani* and placed, with some doubt, in the modern family Eupodidae. They were restudied by Dubinin (1962), who separated them into five species belonging to four modern families: *Protacarus crani* (Pachygnathidae), *Protospeleorchestes pseudoprotacarus* (Nanorchestidae), *Pseudoprotacarus scoticus* (Alicorhagiidae) and *Paraprotacarus hirsti* and *Palaeotydeus devonicus* (Tydeidae). John Kethley, of the Field Museum of Natural History in Chicago, restudied the specimens more recently and considered all to belong to the family Pachygnathidae except the nanorchestid (a family which is nevertheless closely related to the Pachygnathidae in the superfamily Pachygnathoidea). These mites were probably saprophagous, feeding on dead organic matter in litter and soil, though it is possible that some sucked juices from living plants.

Collembolans. Springtails are tiny, jumping animals common just about everywhere. *Rhyniella praecursor* (**93**) was described from the Rhynie Chert by Hirst and Maulik (1926). Various people commented on the identity of *Rhyniella*, but it was not until the 1980s, when Whalley and Jarzembowski ground one of the original specimens down to reveal the entire abdomen with its tail-spring (furcula), that *Rhyniella* could confidently be assigned to the Isotomidae, a family which includes the Glacier and Firn Fleas and other inhabitants of stressed environments.

Euthycarcinoids (**94**). These are an extinct group of arthropods with a fossil record which started in the late Silurian and continued to the mid-Triassic. The first and last records are from Australia, but euthycarcinoids have also been recorded from the Upper Carboniferous of Mazon Creek, USA (Chapter 6), and Europe. *Heterocrania rhyniensis* was described by Hirst and Maulik (1926) as a chelicerate, but the recent discovery of more complete specimens in the Windyfield Chert by Anderson and Trewin revealed that the animal was a euthycarcinoid. It is the first euthycarcinoid known from Devonian rocks. The systematic placement of

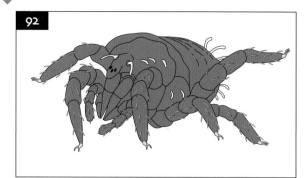

92 Drawing of a Rhynie mite. Body length 0.3 mm (0.01 in) (after Hirst, 1923).

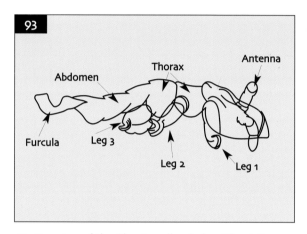

93 Drawing of the Rhynie collembolan *Rhyniella*. Body length 1 mm (0.04 in) (after Whalley and Jarzembowski, 1981).

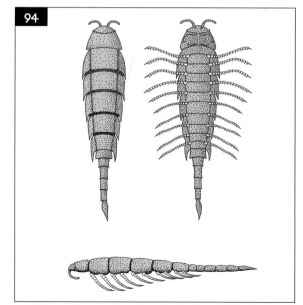

94 Reconstruction of a euthycarcinoid.

euthycarcinoids is problematic because they show similarities to both crustaceans and insects. Euthy-carcinoids were probably detritus-feeders living in ephemeral freshwater pools.

Myriapods. Two myriapods are known from the Rhynie Chert. *Crussolum* is a centipede similar to the house centipede (*Scutigera*) which is common throughout the world's warm climates. It was a fast-running predator with poison jaws. *Leverhulmia mariae* is a small myriapod of uncertain affinity whose main claim to fame is that its gut contents are preserved. It was probably a detritus-feeder living in plant litter which was washed into an ephemeral pool (because the fossil was found among specimens of the aquatic crustacean *Lepidocaris*).

Puncture wounds in some of the Rhynie plants were put forward by Kevan *et al.* (1975) as evidence that some of the animals, possibly the mites, were sucking plant juices. However, most of the animals found in the Rhynie Chert were predators (trigonotarbids and centipedes), whose mouthparts are clearly designed for predation and whose modern descendants are all predatory; or detritivores (*Leverhulmia, Heterocrania*, and the crustacean *Lepidocaris*). Evidence for detritivory comes from the gut content of spores and plant debris in *Leverhulmia*; coprolites (fossil faeces) found in close association with these animals, which may have been produced by them; and the feeding habits of their modern counterparts.

Palaeoecology of the Rhynie Chert

The Rhynie Chert was the first early terrestrial eco-system to be described; it is still the best known and is, surprisingly, still turning up new plants, animals and other information. It represents a hot-spring deposit and so the suggestion that it might preserve a very specialized biota has been discussed. However, we can see from the biota of other early terrestrial sites that they preserve a similar range of plants and animals, and we can envisage an early terrestrial biota which was distinct from later terrestrial biotas. One might expect there to be differences between the early terrestrial biotas because of differences in their preservational setting. Indeed there are, but the differences are primarily between the non-terrestrial components; what is striking is the similarity between the terrestrial plants and animals from these diverse habitats. While earlier phases of plant colonization of land can be recognized (Edwards and Selden, 1993), by Rhynie Chert times there was a considerable diversity in plant morphology, including the first true vascular plants. These provided a short turf, generally less than 200 mm (8 in) in height, on and amongst which the early land animals lived.

The animals show a preponderance of carnivores, some detritivores, but an absence of herbivores. The evidence points strongly towards a food chain based on detritivory, as is common in soils today, in which plant material is first partly decomposed by bacteria and fungi before being ingested by detritivorous animals such as arthropleurids. True herbivores use symbiotic bacteria and fungi in their guts to enable them to digest plant tissues, in effect by-passing the detritivores, and returning material to the soil in the form of faeces. At Rhynie and other early terrestrial sites, either we are sampling a soil ecosystem or the food chain was detritivore-based. There is some evidence for damage to plants by piercing mouthparts at Rhynie (although the perpetrators of the damage are unknown), but conclusive evidence for animals eating the vegetative parts of plants does not appear until the late Carboniferous. It seems that Rhynie, and other Siluro-Devonian terrestrial sites discussed here, preserve a unique type of trophic system which pre-dates the evolution of herbivory, which so dominates the world's ecosystems today (Shear and Selden, 2001).

Comparison of the Rhynie Chert with other early terrestrial biotas

Fifty years passed between the first announcements of the Rhynie Chert early terrestrial plants and animals and the next major discovery of another Devonian land biota, which came in the 1970s with the discovery of lycopsids, rhyniopsids, trigonotarbids and arthro-pleurids (a group of Palaeozoic myriapods) together with amphibious eurypterids and aquatic xiphosurans, crustaceans, molluscs and fish at Alken-an-der-Mosel, Germany (Størmer, 1976). Also in the early 1970s, palaeobotanists Grierson and Bonamo of the State University of New York at Binghamton were extracting fossil plants from approximately 380 Ma Devonian shales at a site near Gilboa, New York, by dissolving the rock in hydrofluoric acid. Amongst the plant cuticles were some remains of animals – the oldest known land animals in North America. Over the next three decades, scientists from the USA and Britain studied the animals, which include trigonotarbids, mites, centipedes, arthropleurids, scorpions, eurypterids, and possible early insects.

In 1990 a site known for its preservation of mixed brackish-water animals (e.g. fish scales) and terrestrial plants at Ludford Lane, Ludlow, Shropshire, UK, dating to about 415 Ma (late Silurian), was examined by a technique similar to that used on the Gilboa material. Though these remains were less well preserved, scientists at Manchester University, UK, extracted trigonotarbids, centipedes, arthropleurids, eurypterids, and scorpions from this locality, giving it the record for the oldest known terrestrial animals anywhere in the world (Jeram *et al.*, 1990). Additional sites have also been found in Canada and, though few specimens have been recovered, they follow the same pattern of trigonotarbids, arthropleurids, and so on.

Further Reading

Anderson, L. I. and Trewin, N. H. 2003. An Early Devonian arthropod fauna from the Windyfield cherts, Aberdeenshire, Scotland. *Palaeontology* **46**, 467–509.

Claridge, M. F. and Lyon, A. G. 1961. Book-lungs in the Devonian Palaeocharinidae (Arachnida). *Nature* **191**, 1190–1191.

Dubinin, V. B. 1962. Class Acaromorpha: mites or gnathosomic chelicerate arthropods? 447–473. *In* Rodendorf, B. B. (ed.). *Fundamentals of Palaeontology Volume 9.* Academy of Sciences of the USSR, Moscow, xxxi + 894 pp.

Edwards, D. and Selden, P. A. 1993. The development of early terrestrial ecosystems. *Botanical Journal of Scotland* **46**, 337–366.

Hirst, S. 1923. On some arachnid remains from the Old Red Sandstone (Rhynie Chert Bed, Aberdeenshire). *Annals and Magazine of Natural History 9th Series* **70**, 455–474.

Hirst, S. and Maulik, S. 1926. On some arthropod remains from the Rhynie Chert (Old Red Sandstone). *Geological Magazine* **63**, 69–71.

Jeram, A. J., Selden, P. A. and Edwards, D. 1990. Land animals in the Silurian: arachnids and myriapods from Shropshire, England. *Science* **250**, 658–661.

Kevan, P. G., Chaloner, W. G. and Savile, D. B. O. 1975. Interrelationships of early terrestrial arthropods and plants. *Palaeontology* **18**, 391–417.

Kidston, R. and Lang, W. H. 1917. On Old Red Sandstone plants showing structure, from the Rhynie Chert bed, Aberdeenshire. Part I. *Rhynia gwynne-vaughani* Kidston and Lang. *Transactions of the Royal Society of Edinburgh* **51**, 761–784.

Kidston, R. and Lang, W. H. 1920. On Old Red Sandstone plants showing structure, from the Rhynie Chert bed, Aberdeenshire. Part II. Additional notes on *Rhynia gwynne-vaughani*, Kidston and Lang; with descriptions of *Rhynia major*, n.sp., and *Hornia lignieri*, n. g., n. sp. *Transactions of the Royal Society of Edinburgh* **52**, 603–627.

Kidston, R. and Lang, W. H. 1920. On Old Red Sandstone plants showing structure, from the Rhynie Chert bed, Aberdeenshire. Part III. *Asteroxlon mackiei*, Kidston and Lang. *Transactions of the Royal Society of Edinburgh*, **52**, 643–680.

Kidston, R. and Lang, W. H. 1921a. On Old Red Sandstone plants showing structure, from the Rhynie Chert bed, Aberdeenshire. Part IV. Restorations of the vascular cryptogams, and discussion of their bearing on the general morphology of the Pteridophyta and the origin of the organisation of land-plants. *Transactions of the Royal Society of Edinburgh* **52**, 831–854.

Kidston, R. and Lang, W. H. 1921b. On Old Red Sandstone plants showing structure, from the Rhynie Chert bed, Aberdeenshire. Part V. The Thallophyta occuring in the peat-bed; the succession of the plants throughout a vertical section of the bed, and the conditions of accumulation and preservation of the deposit. *Transactions of the Royal Society of Edinburgh* **52**, 855–902.

Remy, W., Selden, P. A. and Trewin, N. H. 1999. Gli strati di Rhynie. 28–35. *In* Pinna, G. (ed.). *Alle radici della storia naturale d'Europa.* Jaca Book, Milan, 254 pp.

Remy, W., Selden, P. A. and Trewin, N. H. 2000. Der Rhynie Chert, Unter-Devon, Schottland. 28–35. *In* Pinna, G. and Meischner, D. (eds.). *Europäische Fossillagerstätten.* Springer, Berlin, 264 pp.

Rice, C. M., Trewin, N. H. and Anderson, L. I. 2002. Geological setting of the Early Devonian Rhynie cherts, Aberdeenshire, Scotland: an early terrestrial hot spring system. *Journal of the Geological Society of London* **159**, 203–214.

Rolfe, W. D. I. 1980. Early invertebrate terrestrial faunas. 117–157. *In* Panchen, A. L. (ed.). *The terrestrial environment and the origin of land vertebrates.* Systematics Association Special Volume **15**. Academic Press, London and New York, 633 pp.

Scourfield, D. J. 1926. On a new type of crustacean from the old Red Sandstone (Rhynie Chert Bed, Aberdeenshire) – *Lepidocaris rhyniensis*, gen. et sp. nov. *Philosophical Transactions of the Royal Society of London*, Series B **214**, 153–187.

Scourfield, D. J. 1940. Two new and nearly complete specimens of young stages of the Devonian fossil crustacean *Lepidocaris rhyniensis. Proceedings of the Linnean Society* **152**, 290–298.

Selden, P. A. and Edwards, D. 1989. Colonisation of the land. 122–152. *In* Allen, K. C. and Briggs, D. E. G. (eds.). *Evolution and the fossil record.* Belhaven Press, London, xiii + 265 pp.

Selden, P. A., Shear, W. A. and Bonamo, P. M. 1991. A spider and other arachnids from the Devonian of New York, and reinterpretations of Devonian Araneae. *Palaeontology* **34**, 241–281.

Shear, W. A. 1991. The early development of terrestrial ecosystems. *Nature* **351**, 283–289.

Shear, W. A. and Selden, P. A. 2001. Rustling in the undergrowth: animals in early terrestrial ecosystems. 29–51. *In* Gensel, P. G. and Edwards, D. (eds.). *Plants invade the land: evolutionary and environmental perspectives.* Columbia University Press, New York, x + 304 pp.

Størmer, L. 1976. Arthropods from the Lower Devonian (Lower Emsian) of Alken-an-der-Mosel, Germany. Part 5. Myriapoda and additional forms, with general remarks on fauna and problems regarding invasion of land by arthropods. *Senckenbergiana Lethaea* **57**, 87–183.

Trewin, N. H. 1994. Depositional environment and preservation of biota in the Lower Devonian hot-springs of Rhynie, Aberdeenshire, Scotland. *Transactions of the Royal Society of Edinburgh: Earth Sciences* **84**, 433–442.

Whalley, P. E. and Jarzembowski, E. A. 1981. A new assessment of *Rhyniella*, the earliest known insect, from the Devonian of Rhynie, Scotland. *Nature* **291**, 317.

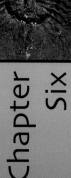

MAZON CREEK

Chapter Six

Background: the Coal Measures

Once life became established on land, it quickly developed a more complex habitat structure; as plants developed tree forms so forests evolved (by the Late Devonian), and alongside this there was a great diversification of animal life. Early Carboniferous times saw the evolution of tetrapods – four-legged vertebrates – which emerged onto land to start preying on the abundant invertebrate life already established there. By the Late Carboniferous there were extensive forests across the equatorial regions of the globe, which included the present areas occupied by north-west and central Europe, eastern and central USA, and elsewhere in the world such as southern China and South America. These forests are represented in the fossil record by coal seams, which are preserved because in these places the forests developed on mires – permanently waterlogged ground – the anoxic conditions of which prevented complete decay of the forest litter and thus allowed the formation of peat which, when compressed under a great thickness of later sediment, turned to coal. It was these vast beds of coal, and associated ironstones, clays, and other natural resources, which in Britain provided the raw materials for the Industrial Revolution.

A Coal Measure sequence of rocks presents a more complex and interesting suite of environments than a simple swamp forest, and most Upper Carboniferous Coal Measure sequences represent deltas with a range of environments from marine bays through brackish lagoons to sand bars, freshwater lakes, levees and swamp forests. Delta lobes are geologically short-lived. If their sediment supply is cut off, they rapidly sink and sea water transgresses the land surface, swamping the forests. Thus, in many sequences there is a band of mud bearing marine fossils immediately above a coal seam. The sea may persist in the area for many tens or hundreds of years before a new delta lobe builds out into the area, and relatively quickly establishes new silt and sand substrates upon which new forests can develop. Even in established swamp forests, floods are common-place. As a result of these changing environments, Coal Measure sequences show a distinctive pattern of thin mud or shale layers, siltstones (often regularly laminated), coarse sandstones, and coal seams.

Some of the vascular plant groups discussed in Chapter 5 continued with little morphological change into the Carboniferous and beyond, for example the bryophytes, while others, such as the psilophytes, gave rise to the horsetails, clubmosses and ferns, which attained gigantic proportions in the Carboniferous. Many of these groups also formed the understorey vegetation, together with seed-ferns, cordaites and early conifers. Animals, too, had diversified and moved into the new niches provided by the plants. Insects had appeared, evolved wings, and some Carboniferous early dragonflies had a wingspan of 0.75 m (2.5 ft) (though those found in the Mazon Creek beds were smaller). Myriapods became giant in the Carboniferous too, with large, armoured millipedes and the enormous (more than 2 m [6 ft] long) arthropleurids – the largest known land arthropods. Vertebrates had not only followed the arthropods onto land but amphibious tetrapods had also attained large size (up to 1 m [3 ft] long), and there were freshwater sharks with bizarre dorsal spines.

Mazon Creek is a small tributary of the Illinois River, situated some 150 km (90 miles) south-west of Chicago (**95**), and has given its name to this Lagerstätte, which actually comes principally from spoil heaps of the strip coal mines which have operated in the area over the last century. The importance of the Mazon Creek biota is that it has been so well collected, particularly by an army of keen amateurs, that it has yielded the most complete record of late Palaeozoic shallow marine, freshwater and terrestrial life. More than 200 species of plants and 300 animal species have been described, including representatives of 11 animal phyla.

History of discovery of the Mazon Creek biota

Plant fossils were collected and described from natural outcrops and small mine tips in the Mazon Creek area for many years before the large Pit 11 open-strip mine was opened in the 1950s. In the late years of that decade, Peabody Coal bought out the Northern Illinois Coal Company and allowed local fossil collectors to visit the pit and collect from the waste material. In strip mining, the overburden (in this case the Francis Creek Shale) is stripped off by giant buckets on drag lines to reveal the coal beneath, which is then simply dug out by smaller diggers and loaded into trucks for removal to the sorting plant. The coal is dug in long strips, so the overburden is used to back-fill the strip where coal was previously removed. It is the Francis Creek Shale which is the source of the exceptional fossils at Mazon Creek. Once news of the coal company's generosity in allowing access to its site spread, there was a regular stream of amateur fossil collectors to the site looking for the elusive special fossil.

The fossils at Mazon Creek occur in clay ironstone nodules (concretions). These usually require a winter or so of weathering before they will split easily with a single hammer-blow, usually along the weakest line – which is that of the fossil. Some collectors artificially freeze and thaw the nodules to accelerate this process. Many of the concretions contain seed-fern fronds, some contain indeterminate shapes which were termed 'blobs' and consequently thrown away. Later research has shown that most of these blobs are actually fossil jellyfish, attesting to the unusual preservation of soft-bodied animals at Mazon Creek. The scramble of collectors at Pit 11 could have resulted in the loss of the best fossils to personal cabinets, never to be studied by experts were it not for the efforts of Dr E. S. ('Gene') Richardson, who encouraged the collectors to meet at regular intervals at the Field Museum in Chicago and show their finds. In this way they learned from both the scientists and each other about what the animals and plants were and the best ways of finding them. They could be swapped, and the best specimens presented to the museum for study. The annual Open House for Mazon Creek fossils at the Field Museum continues to this day.

Some 20 years ago the Peabody Coal Company sold Pit 11 for the construction of a nuclear power plant. While mining no longer goes on there, the tips remain and are still picked over. Moreover, during construction of the power station, boreholes were drilled which passed through the Francis Creek Shale and yielded some important information about its environment of deposition.

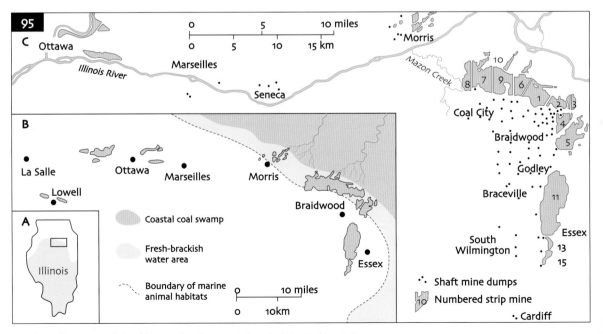

95 Locality map of the Mazon Creek area (after Baird *et al.*, 1986).

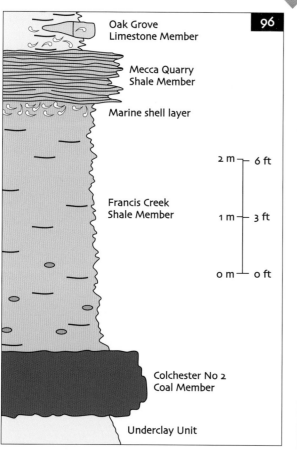

96 Stratigraphic log of the Francis Creek Shale and associated Members of the Carbondale Formation, Illinois (after Baird *et al.*, 1986).

97 Disused Pit 8 in the Colchester Nº 2 Coal. The spoil forming the raised banks is the source of the Mazon Creek nodules.

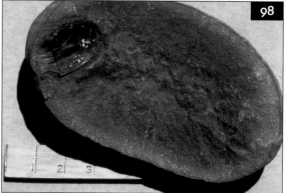

98 Death trail of undescribed solemyid bivalve. The bivalve was still alive and attempting to escape while the siderite nodule was forming, resulting in a fossilized trail and the bivalve at the edge of the nodule (CFM). Scale bar is in cm.

Stratigraphic setting and taphonomy of the Mazon Creek biota

The Mazon Creek fossils occur in siderite (ironstone, $FeCO_3$) concretions in the Francis Creek Shale Member of the Carbondale Formation, of Westphalian D age. The Francis Creek Shale overlies the Colchester Nº 2 Coal Member, and is itself overlain by the Mecca Quarry Shale Member (**96, 97**). The Colchester Coal is generally about 1 m (3 ft) thick; the Francis Creek Shale is typically a grey, muddy siltstone with minor sandstones and varies from complete absence up to 25 m (80 ft) or more in thickness. The siderite concretions occur only in the lower 3–5 m (10–15 ft) of the member, and only where the shale is more than 15 m (50 ft) thick. The shale is coarser and bears sandstones near its top. The Mecca Quarry Shale is a typical Pennsylvanian black shale, which peels easily into sheets, and contains a rich fauna of sharks and their coprolites, which was exhaustively monographed by Zangerl and Richardson (1963). It is generally about 0.5 m (1 ft 6 in) thick but is absent over areas where the Francis Creek Shale is more than about 10 m (30 ft) thick, and thus also over the nodule-bearing parts of the Francis Creek Shale.

The fossils are found almost only within the siderite concretions. When these are broken open they reveal a nearly three-dimensionally preserved organism, though the fossil becomes more flattened towards the edge of the nodule. Fossils are generally preserved as external moulds, commonly with a carbonaceous film (if a plant). There may be crystals such as pyrite, calcite or sphalerite on the mould surfaces. The commonest mineral on these surfaces, however, is kaolinite – a white, soapy, clay mineral. This is often found completely infilling the space between the moulds (i.e. forms a cast). It is soft and therefore easy to remove mechanically. Fossils with few hard parts or little rigidity, such as jellyfish, may collapse completely and be preserved as composite moulds; even arthropods may show dorsal and ventral structures superimposed.

Most fossils show very little decay and, indeed there are instances of bivalves preserved on the edge of the nodule at the end of their death trail! (**98**). The concretions do not normally extend far beyond the fossil, so the size and shape of the nodule correlates with the organism inside. Few concretions exceed 300 mm (12 in), so large animals are rare in the biota. The evidence presented so far tends to indicate that the

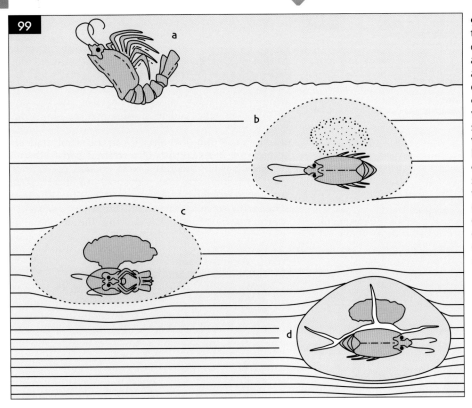

99   Diagram showing the rapid formation of a siderite nodule around a dead shrimp. a, Dead shrimp lands on sea floor; b, partial decay by bacteria, volatiles rise; c, siderite precipitation while compaction begins; d, further compaction of surrounding sediment while dewatering (syneresis) causes cracks to extend from centre of nodule outwards (after Baird *et al.*, 1986).

concretions formed very soon after the death and burial of the organisms – molluscs stopped in their tracks, seed-fern pinnules at right-angles to the bedding, and little decay (**99**). Large fish and amphibians, it is presumed, could escape this environment and not be preserved. That the organisms are preserved three-dimensionally, at least in the cores of the nodules, while the surrounding matrix, like all siltstones, is greatly compressed implies that the concretions formed before any appreciable compaction. Indeed, the siltstone laminae can be seen (**100**) to widen progressively from the matrix towards the centre of the nodules, suggesting that the concretions grew during compaction. Also, cracks within the nodules, commonly infilled with kaolinite, can be related to dewatering of the sediment (syneresis) during their formation.

Because the fossil-bearing concretions mirror the shape of the fossil, and the fossils are generally situated fairly centrally within the concretions, we can assume that the organisms contributed significantly to their formation. Moreover, barren nodules can usually be explained as containing rather flimsy fossils, unrecognizable organic matter, trace fossils, and the like. The nodules contain about 80% siderite cement, which implies that when the concretions formed there was at least 80% water by volume in the sediment before compaction. Iron would normally react with sulphur under the influence of anaerobic bacteria in the presence of decaying organic matter to form pyrite (FeS_2), as in the Hunsrück Slate (Chapter 4), in preference to siderite, but once any sulphate was used up by this process (some pyrite does occur in the nodules), then methanogenic bacteria would help to generate

100   Section through the Francis Creek Shale laminites and part of a siderite nodule. Note the laminae widen towards the nodule, indicating greater compaction of the surrounding silt/clay laminites than the nodule, and that compaction had started during growth of the nodule. Thickest part of specimen (left) is 4 cm (1.6 in).

siderite. Conditions at Mazon Creek which helped this process would have been an abundance of iron and a weak supply of sulphate. The Mazon Creek nodules are commonly asymmetrical, with flatter bottoms and more pointed tops. This is due to the effects of gravity: the weight of the carcass presses into the sediment beneath, the concretion can grow more easily upwards (where there is less compaction), and any light fluids resulting from decay would also rise preferentially.

Description of the Mazon Creek biota

The Mazon Creek biota actually consists of two biotas: the Braidwood and the Essex; the former occurs mainly in the north of the area, the latter in the south. Eighty-three percent of the Braidwood nodules contain plants, with the next commonest (7.8%) inclusions being coprolites. Following these (in descending order) can be found: freshwater bivalves (1.8%), freshwater shrimps (0.5%), other molluscs (0.4%), horseshoe crabs (0.3%), millipedes (0.1%), fish scales (0.1%), then insects, arachnids, fish and centipedes form the remainder (< 0.1%). On the other hand, only 29% of the Essex biota consists of plants, the commonest animal being the 'blob' *Essexella* (42%). Following these (in descending order) are burrows and trails (5.9%), the marine solemyid bivalves (5.5%), coprolites (4.8%), worms (2.8%), miscellaneous molluscs (1.9%), the marine shrimp *Belotelson* (1.8%), the marine bivalve *Myalinella* (1.4%), miscellaneous shrimps (0.5%), the crustacean *Cyclus* (0.5%), the enigmatic Tully Monster (0.4%), the scallop *Pecten* (0.3%), the jellyfish *Octomedusa* (0.3%), miscellaneous fish (0.2%), then insects, millipedes and centipedes, hydroids, horseshoe crabs, arachnids, and amphibians form the remainder (< 0.1%). From this list it can be seen that the Braidwood biota consists of terrestrial and freshwater organisms while the Essex biota consists of predominantly marine organisms, with some drifted plants and so on. No marine organisms can drift into fresh water but freshwater and terrestrial organisms can be washed down into the sea. Notice, too, that the marine Essex biota is not a typical marine biota (there are no brachiopods, corals, crinoids, and so on), so it must represent a reduced-salinity, and perhaps muddy, environment which fully marine animals cannot tolerate.

Plants. The Mazon Creek nodules preserve a typical Coal Measure flora, exceptional only in that its preservation is better in the concretions than in other Upper Carboniferous clayrocks. Large fossils are not preserved in the Francis Creek Shale, but the presence of tree-sized clubmosses and horsetails can be inferred from pieces of bark (*Lepidodendron* and *Calamites*, respectively) and foliage (*Lepidophylloides* and *Annularia*, respectively). Seed-fern pinnules are the commonest fossils in all Mazon Creek nodules; examples include *Neuropteris* (**101**), *Pecopteris* (**102**), and *Alethopteris* (**103**). Comparison with modern coastal swamp forests suggests that much of the plant debris found in the Mazon Creek nodules originated not from the forest itself but that it was carried down by streams from far inland, so possibly represents a mixture of coastal and upland forest. Plant debris is allochthonous (i.e. drifted from its original life position) in both Braidwood and Essex biotas, but is more common in the former.

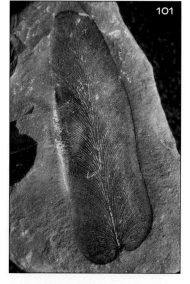

101 The seed-fern *Neuropteris* (MU). Pinnule is 6 cm (2.4 in) long.

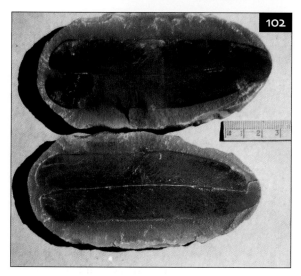

102 The seed-fern *Pecopteris* (MU). Scale bar is in cm.

103 The seed-fern *Alethopteris* (MU). Nodule is 7 cm (2.7 in) long.

Cnidarians. This phylum includes the corals and anemones (Anthozoa), hydroids (Hydrozoa), Scyphozoa and Cubozoa (jellyfish), and a few other groups. Apart from the corals, they have no mineralized skeletons but some jellyfish have stiffened parts. Most 'blobs' in Mazon Creek nodules are jellyfish remains, and some show distinct tentacles and other structures. *Essexella*, for example (**104**), shows a bell with a cylindrical sheet hanging below it. *Essexella* belongs in the Scyphozoa, while *Anthracomedusa* (**105**), with four bunches of numerous tentacles, is a cubozoan.

Bivalves. Bivalves are common Carboniferous fossils on account of their hard, calcareous shells (valves), but those in the Mazon Creek nodules are important because they commonly preserve soft-part morphology. There is a high diversity of bivalves in the Mazon Creek, with 12 superfamilies represented. It is convenient to divide the fauna into freshwater and marine forms: i.e. those found in the Braidwood and Essex biotas respectively. The most common marine bivalve occurs as articulated valves (called 'clam-clam' specimens by amateur collectors), and is occasionally found at the end of an unsuccessful escape trail (**98**). Previously misidentified as *Edmondia*, it has been shown recently to be an undescribed solemyid; true *Edmondia* are rare at Mazon Creek. These solemyids are burrowers (infaunal benthos), but other common bivalves in the Essex biota are the thin-shelled swimmers (nekton) *Myalinella* and *Aviculopecten* (**106**). The family Myalinidae also includes freshwater forms such as *Anthraconaia*, found in the Braidwood biota.

Other molluscs. Three other classes of mollusc occur in the Mazon Creek biota: Polyplacophora, Gastropoda and Cephalopoda. Though there are many freshwater gastropods (snails) today, the only gastropods at Mazon Creek are from the marine Essex biota. Polyplacophora (chitons) are exclusively marine animals which are rare in the fossil record because they usually inhabit rocky shores. However, one genus, *Glaphurochiton* (**107**), occurs in the Essex biota. Cephalopods are also wholly marine and, though normally common in fully marine Carboniferous rocks, they are rarer than chitons at Mazon Creek, but quite diverse in the Essex biota. In addition to the orthocone bactritoids, coiled ammonoids and nautiloids with external shells, coleoids (with internal hard parts) are also present. *Jeletzkya* is a small, squid-like coleoid with an internal shell similar to a cuttlebone.

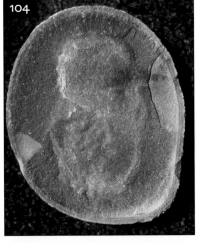

104 The jellyfish *Essexella asherae* (MU). Nodule is 6 cm (2.4 in) long.

105 The jellyfish *Anthracomedusa turnbulli* (MU). Scale bar is in cm.

106 The bivalve *Aviculopecten mazonensis* (MU). Scale bar is in cm.

Worms. Apart from their tiny jaws, called scolecodonts, polychaete annelids are rarely preserved as fossils because they are soft-bodied. Therefore, the great diversity of polychaetes found at Mazon Creek is an important contribution to the fossil record of this important group of marine animals. One of the commonest Mazon Creek polychaetes is *Astreptoscolex* (**108**), which shows a segmented body and short chaetae (spines) along each side.

Shrimps. A wide variety of Crustacea occur at Mazon Creek, many of which have a shrimp-like body shape. There are both freshwater and marine shrimps, found in the Braidwood and Essex biotas, respectively. *Belotelson magister*, a robust species, is by far the most common shrimp in the marine Essex biota. The second commonest marine shrimp is *Kallidecthes. Acanthotelson* (**109**) and *Palaeocaris* are freshwater shrimps which occur in the Braidwood biota and are also found rarely in the Essex nodules, where they presumably were washed down by currents.

Other crustaceans. Some squat, crayfish-like forms also occur at Mazon Creek: *Anthracaris* in the Braidwood

biota and *Mamayocaris* in the Essex biota. There are also phyllocarids (*Dithyrocaris*) in the Essex biota, conchostracans (crustaceans almost completely enclosed in a bivalved carapace) which were probably fresh- or brackish-water forms, some ostracodes and barnacles. A crustacean which commonly occurs in Upper Carboniferous nodules, *Cyclus*, is also found in the Essex biota. As its name suggests, *Cyclus* has a round, dish-like carapace, and has been thought of as a fish parasite, though it may have been free-living.

Chelicerates. The arthropod subphylum Chelicerata includes horseshoe crabs (Xiphosura), scorpions, eurypterids, spiders, mites and other arachnids. Mazon Creek has yielded some exceptionally fine fossils of these animals, which have provided a great deal of information on the evolutionary history of the chelicerates. *Euproops danae* (**110**) is one of the best known horseshoe crabs in the fossil record. Work by Dan Fisher (1979) of the University of Michigan revealed the amphibious mode of life of these animals, which are found predominantly in the Braidwood biota.

107 The chiton *Glaphurochiton concinnus* (MU). Scale bar is in cm.

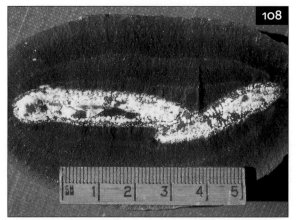

108 The polychate annelid *Astreptoscolex anasillosus* (MU). Scale bar is in cm.

109 The shrimp *Acanthotelson stimpsoni* (MU). Scale bar is in cm.

110 The horseshoe crab *Euproops danae* (MU). Scale bar is in cm.

Eurypterids were discussed in Chapter 3 in connection with the Soom Shale. By the Upper Carboniferous they were mostly amphibious and represented largely by the genus *Adelophthalmus*, with numerous specimens in Mazon Creek nodules.

Terrestrial arachnids are well represented in the Braidwood biota, and the extinct group Phalangiotarbida has the most numerous representatives; trigonotarbids are also well represented. The latter are close to spiders in morphology but lack poison glands and silk and, as arachnids go, are relatively common in Upper Palaeozoic terrestrial ecosystems. Two living orders of arachnids, Uropygi (whip scorpions) and Amblypygi (whip spiders), have some well-preserved examples in the Braidwood biota (**111**). An interesting order of living arachnids, Ricinulei, is represented at Mazon Creek by three genera. Ricinulei are rarely

encountered, even today, and are restricted to tropical forests and caves. They are only known from the Upper Carboniferous and Recent, but seem to have changed little in between. Three other arachnid orders occur at Mazon Creek: Opilionida (harvestmen), Solpugida (camel spiders) and Scorpionida. The scorpions are the most ancient of arachnids and, though exclusively terrestrial today, are found in aquatic environments in the Silurian. Scorpions are found only in terrestrial environments by the Upper Carboniferous, and are the second most abundant arachnid in the Braidwood biota.

Insects. Six orders of insects (only one of which, Blattodea – cockroaches – is still extant) occur at Mazon Creek. Much of our knowledge of Upper Carboniferous insects comes from the 150 species found in Mazon Creek nodules. Palaeodictyoptera were medium to large flying forms with patterned wings. Both nymphs and adults were terrestrial and had sucking mouthparts. Megasecoptera were similar to Palaeodictyoptera but had more slender, often petiolated (stalked) wings. Diaphanopterodea resembled Megasecoptera with the one exception that they could fold their wings over their backs as do modern butterflies and damselflies. Protodonata, as their name suggests, were closely related to modern Odonata (dragonflies and damselflies). Some reached giant proportions in the Upper Carboniferous. It is presumed that, like Odonata, the nymphs led an aquatic life, but none have been found. Protorthoptera were related to the modern Orthoptera (grasshoppers and allies) but lacked jumping legs. Protorthoptera is the largest order of extinct insects, with some 50 families known from the Upper Carboniferous and Permian. Twelve families occur at Mazon Creek, and *Gerarus* is the commonest insect fossil found there. Blattodea (cockroaches) (**112**) are the most numerous insects in Upper Carboniferous nodules but are not so abundant at Mazon Creek; their veined wings are often mistaken for seed-fern pinnules.

Myriapods. Myriapods are multilegged arthropods which include the centipedes (Chilopoda), millipedes (Diplopoda), two other living classes (Symphyla and Pauropoda), and the Palaeozoic Arthropleurida. Myriapods were among the earliest known land animals (Chapter 5), but by Upper Carboniferous times some had become gigantic, and many sported fierce spines, presumably for defence against predators. Short millipedes which could roll up into a ball (Oniscomorpha: *Amynilyspes*) were present in the Braidwood biota, but the most dramatic forms belong to the extinct order Euphoberiida. *Myriacantherpestes* (**113**) probably reached 300 mm in length and had long, forked, lateral spines and shorter dorsal spines. *Xyloiulus* (**114**) was a more typical, cylindrical spirobolid millipede. Millipedes are generally detritus feeders, while centipedes are carnivorous. Centipedes at Mazon Creek include the scolopendromorph *Mazoscolopendra* and the fast-running scutigeromorph *Latzelia*. Arthropleurids range from tiny forms in the Silurian to the Upper Carboniferous, when they became the largest known terrestrial animals, reaching 2 m (6–7 ft) in length. Nevertheless, like millipedes, they were probably detritus-feeders. Isolated legs and plates of *Arthropleura* occur at Mazon Creek.

111 Cast of uropygid arachnid *Geralinura carbonaria* (CFM). Nodule is 6 cm (2.4 in) long.

112 Cockroach (Blattodea) (MU). Scale bar is in cm.

Onychophora (velvet worms) should also be mentioned here. They are known from the Cambrian (e.g. *Aysheaia*, Chapter 2), when they were marine, to the Recent, when they are wholly terrestrial. The Mazon Creek *Ilyodes* was collected from natural outcrops and it is not known whether it came from the terrestrial Braidwood or marine Essex biota.

Other arthropods. Euthycarcinoids are an odd group of apparently uniramous arthropods (those with a single leg-branch, like myriapods) which range from the Silurian to Triassic. Mazon Creek has three species.

Another group of arthropods of unknown affinity is the Thylacocephala, which ranged from Cambrian to Cretaceous. Commonly called flea-shrimps, they may or may not belong to the Crustacea. They have a bivalved carapace which encloses most of the body, and large eyes. *Concavicaris* is quite common in Essex nodules.

Other invertebrates. Brachiopods are common in marine sediments of normal salinity, but they are rare in Mazon Creek nodules, being represented only by inarticulates such as *Lingula*, which is known to prefer brackish waters. *Lingula* is the only infaunal brachiopod, living in a vertical burrow into which it can retract by means of a long, fleshy pedicle. Numerous specimens from Pit 11 preserve *Lingula* in life position with burrow and pedicle intact.

Like brachiopods, echinoderms are usually found in fully marine waters and, apart from one crinoid specimen, the only echinoderm found at Mazon Creek is the holothurian (sea cucumber) *Achistrum*, which is actually quite common in Essex nodules. It can be distinguished from other worm-like creatures by the ring of calcareous plates which forms part of the sphincter at one end of the animal.

Perhaps the most interesting animal of all at Mazon Creek is one popularly known as 'Tully's Monster', named after its discoverer, the avid collector Francis Tully. *Tullimonstrum gregarium*, to give it its scientific name, ranges up to 300 mm (12 in) long. It has a segmented, sausage-shaped body with a long proboscis at the anterior, which terminates in a claw with up to 14 tiny teeth (**115**). Posteriorly, there is a diamond-shaped tail fin. Near the base of the proboscis is a crescentic structure, and just behind this is a transverse bar bearing an eye at each end. A number of ideas have been put forward as to the affinity of *Tullimonstrum*: conodont animal, annelid, nemertean, mollusc, or a group on its own. *Tullimonstrum* was clearly nektonic and predatory, and its overall appearance, proboscis, eyes and tooth structure are all very reminiscent of a group of shell-less gastropods known as heteropodids. The fame of the Tully Monster was assured a few years ago when it was voted as State Fossil of Illinois.

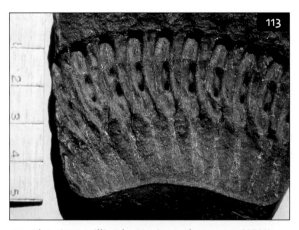

113 The giant millipede *Myriacantherpestes* (CFM). Scale bar is in cm.

114 The millipede *Xyloiulus* (CFM). Nodule is 7 cm (2.7 in) long.

115 The 'Tully Monster', *Tullimonstrum gregarium* (MU). Scale bar is in cm.

Fish. More than 30 species of fish are known from Mazon Creek, but their identification is hampered by the fact that many of the fossils are small juveniles or isolated scales. Agnathans are represented by a hagfish and a lamprey, as well as two agnathans which cannot be assigned to known groups. Cartilaginous fish (chondrichthyans) are represented at Mazon Creek by a rare but diverse fauna of mainly juveniles. Interestingly, they do not appear to be the juveniles of the better known Mecca Quarry Shale sharks described by Zangerl and Richardson (1963), which were approximately coeval with the Francis Creek Shale forms, but seem to be the young of sharks which lived in a different habitat from the Mecca Quarry. *Palaeoxyris*, which is believed to be a shark egg-case, is a common fossil in the Braidwood biota and, to a lesser extent, the Essex biota. About 15 genera of bony fish (osteichthyans) occur at Mazon Creek. Most specimens are small and not easy to identify. However, there is a great variety of types from different habitats, from fresh through brackish to marine waters. Palaeoniscids, including deep-bodied forms as well as fusiform species generally referred to as '*Elonichthys*' are common in both Essex and Braidwood biotas. Among sarcopterygians, rhipidistians (which gave rise to tetrapods), coelacanths and lungfish are all present at Mazon Creek.

Tetrapods. Tetrapods are rare at Mazon Creek, but are diverse and include 23 specimens of amphibian and one reptile. Temnospondyl amphibians are represented by larval *Saurerpeton*, both adult and larval *Amphibamus*, and a possible branchiosaurid. A single fragment (4 vertebrae) of an anthracosaurid is known. Aïstopods – limbless, snake-like amphibians – are represented by two species and numerous specimens, and orders Nectridea, Lysorophia and Microsauria are represented by single specimens. A single, immature specimen of a lizard-like, captorhinomorph reptile is known.

Coprolites. These are fossil faeces, which can occur in both Essex and Braidwood biotas. Though not as aesthetically pleasing as plants or animals, coprolites can tell us a great deal about what animals were eating. For example, spiral coprolites containing fish remains indicate that there were probably quite large sharks swimming in the Mazon Creek area, for which we have no body fossil evidence.

Palaeoecology of the Mazon Creek biota

It is obvious from the evidence presented that the Mazon Creek area represents a variety of habitats – terrestrial, fresh water, brackish, and restricted marine – associated with a deltaic environment. The Colchester Coal represents an environment of swamp forest dominated by tree-sized clubmosses and horsetails with an understorey of seed-ferns, amongst other plants. The Francis Creek flora is dominated by fern, seed-fern and horsetail debris, which suggests it came from a more upland setting. The terrestrial animal fauna, such as myriapods, arachnids and insects, presumably lived amongst these plants.

The Francis Creek Shale coarsens upwards, so the initial inundation of the swamp forest was rapid, and later delta-derived sediments filled the marine embayment. Many sedimentological and palaeontological features suggest rapid deposition: failed bivalve escape structures, *Lingula* buried in life position, and edgewise seed fern pinnules, as well as the preservational features associated with rapid burial mentioned under taphonomy, above. Rapid sedimentation is characteristic of conditions adjacent to a delta. Cores resulting from boreholes drilled to test the foundation of the nuclear power plant in the vicinity of Pit 11 revealed complete sedimentation records. Moreover, the clay–silt laminae in these cores are paired, and the pairs widen and narrow in a cyclical fashion (**116**). Kuecher *et al.* (1990) studied the cores and the cyclicity of the paired laminae and interpreted the cyclicity as tidal in origin. The thin clay bands represent still-stands, either flood slack or ebb slack, i.e. when the tide is in the process of turning and the water is not flowing in either direction (**117**). The thicker silt layers represent periods of greater deposition (from the landward direction). The thicker silt bands were laid down during the ebb tide, when the outgoing tide allowed a high water flow into the basin; the narrower silt bands represent flood tides, when the incoming tide resists the flow to some extent. Thus, a single tidal cycle consists of two clay bands and two silt bands. The widest bands correspond to spring tides, with the highest tidal range, and the narrowest to neaps. Kuecher *et al.* (1990)

116 Cycles of clay–silt pairs in the Francis Creek Shale laminites. (MU). Scale bar is in mm.

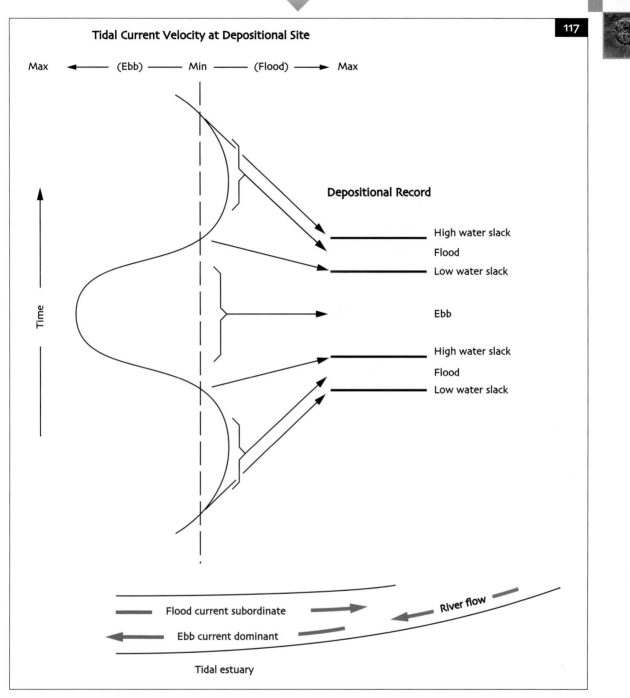

117 Diagram illustrating the formation of cyclical silt–clay pairs by tidal deposition (after Kuecher *et al.*, 1990).

found that there were 15–16 tides in a cycle from springs to next springs. This corresponds to half a lunar month; a complete lunar cycle has two springs (when the Moon and Sun align with the Earth) and two neaps (when the Moon–Earth axis is perpendicular to the Sun–Earth axis). Evidence from the cyclicity of coral growth (Johnson and Nudds, 1975) suggests that the lunar month consisted of 30 days in the Carboniferous period (i.e. the Earth–Moon system has slowed over 350 million years to today's 28-day lunar cycle). Thus, the tidal cycle at Mazon Creek was the diurnal type, as found today in some parts of the world such as the Gulf of Mexico; tides around British coasts are semi-diurnal (i.e. there are two highs and two lows in a 24-hour period).

The tidal cyclicity provides a rare, direct measure of sedimentation rate. Each fortnightly cycle measures from 19 to 85 mm (0.75 to 3.3 in) in thickness. This provides a deposition rate of about 0.5–2.0 m (1.5–6.5 ft) per year of compacted sediment. The entire Francis Creek Shale was therefore deposited in 10–50 years. The tidal cycles provide independent, quantitative evidence of rapid sedimentation already concluded from qualitative evidence from sediments and fossils.

Comparison of Mazon Creek with other Upper Palaeozoic biotas

Calver (1968) recognized a sequence of onshore–offshore communities in the hard-part fossil record of the Westphalian strata of northern England. His estheriid association roughly corresponds to the Braidwood biota, and his myalinid association, consisting principally of *Edmondia* and myalinid bivalves, corresponds to the Essex biota. It appears that the Mazon Creek biotas are common in Upper Carboniferous deltaic settings, but that Mazon Creek exceptionally preserves the soft-bodied biota which is normally lost by taphonomic processes at other localities.

Mazon Creek-type biotas have been found in ironstone concretions in other parts of the world, but nowhere have they been as well studied. For example, a number of localities in the British Coal Measures have yielded good nodule biotas, e.g. Sparth Bottoms (Rochdale), Coseley (West Midlands) and Bickershaw, Lancashire (Anderson *et al.*, 1997). A similar biota occurs at Montceau-les-Mines in France (Poplin and Heyler, 1994). Other Upper Carboniferous localities contribute additional information to our knowledge of Upper Carboniferous non-marine life. For example, the non-nodule locality of Nýřany (Czech Republic) is renowned for its exceptional tetrapod fossils. Schram (1979), in a systematic study of mainly crustaceans in Carboniferous non-marine biotas, argued that stable, predictable associations persisted throughout the period in a continuum. The Grès à Voltzia biota (Chapter 7) is an extension of Schram's continuum into the Triassic (Briggs and Gall, 1990). Thus, the marginal marine ecosystem seems to have been little affected by the great Permo-Triassic extinction event.

Further Reading

Anderson, L. I., Dunlop, J. A., Horrocks, C. A., Winkelmann, H. M. and Eagar, R. M. C. 1997. Exceptionally preserved fossils from Bickershaw, Lancashire, UK (Upper Carboniferous, Westphalian A (Langsettian)). *Geological Journal* **32**, 197–210.

Baird, G. C., Sroka, S. D., Shabica, C. W. and Kuecher, G. J. 1986. Taphonomy of Middle Pennsylvanian Mazon Creek area fossil localities, northeast Illinois: significance of exceptional fossil preservation in syngenetic concretions. *Palaios* **1**, 271–285.

Briggs, D. E. G. and Gall, J.-C. 1990. The continuum in soft-bodied biotas from transitional environments: a quantitative comparison of Triassic and Carboniferous Konservat-Lagerstätten. *Paleobiology* **16**, 204–218.

Calver, M. A. 1968. Distribution of Westphalian marine faunas in northern England and adjoining areas. *Proceedings of the Yorkshire Geological Society* **37**, 1–72.

Johnson, G. A. L. and Nudds, J. R. 1975. Carboniferous coral geochronometers. 27–42. *In* Rosenberg, G. D. and Runcorn, S. K. (eds.). *Growth rhythms and the history of the Earth's rotation.* John Wiley, London, 559pp.

Kuecher, G. J., Woodland, B. G. and Broadhurst, F. M. 1990. Evidence of deposition from individual tides and of tidal cycles from the Francis Creek Shale (host rock to the Mazon Creek Biota), Westphalian D (Pennsylvanian), northeastern Illinois. *Sedimentary Geology* **68**, 211–221.

Nitecki, M. H. (ed.). 1979. *Mazon Creek fossils.* Academic Press, New York, 581pp.

Poplin, C. and Heyler, D. (eds.). 1994. *Quand le Massif Central était sous l'Équateur. Un Écosystème Carbonifère à Montceau-les-Mines.* Comité des Travaux Historiques et Scientifiques, Paris, 341 pp.

Richardson, E. S. and Johnson, R. G. 1971. The Mazon Creek faunas. *Proceedings of the North American Paleontological Convention, 5–7 September 1969, Field Museum of Natural History* **1**, 1222–1235.

Schram, F. R. 1979. The Mazon Creek biotas in the context of a Carboniferous faunal continuum. 159–190. *In* Nitecki, M. H. (ed.). *Mazon Creek fossils.* Academic Press, New York, 581 pp.

Shabica, C. W. and Hay, A. A. (eds.). 1997. *Richardson's guide to the fossil fauna of Mazon Creek.* Northeastern Illinois University, Chicago, xvii + 308 pp.

Zangerl, R. and Richardson, E. S. 1963. The paleoecological history of two Pennsylvanian black shales. *Fieldiana Geology Memoir* **4**, 1–352.

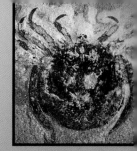

GRÈS À VOLTZIA

Chapter Seven

Background: the Permo-Triassic transition

In the last two chapters, we saw how the terrestrial environment became green with plant life, closely followed by animals rustling through the undergrowth; by the Carboniferous Period lush tropical forests were widespread. These forests were dominated by lycopod trees and pteridosperm (seed-fern) understorey, and animal life included amphibians, insects and arachnids. The Palaeozoic seas had witnessed the rise of arthropods (e.g. trilobites, eurypterids), planktonic graptolites, and brachiopods, and coral reefs were widespread in tropical regions. However, the end of both the Permian Period and the Palaeozoic Era was defined by the sudden change in fauna and flora which resulted from a major mass extinction event – the greatest the Earth has ever witnessed. As we now enter the Triassic Period, and the Mesozoic Era, life on Earth is quite different from what it was before. Trilobites, eurypterids and graptolites are extinct; bivalved molluscs, not brachiopods, are now the dominant shelled animals on the sea floor; there are new types of corals forming reefs; and gymnospermous trees now dominate the flora on land.

The Permian Period is relatively barren of Fossil-Lagerstätten. Nevada has a locality (Buck Mountain) which preserves soft-part morphology of cephalopods, but for terrestrial fossils the Karoo Supergroup of South Africa is justly famous for its preservation of fossil reptiles, especially those which ultimately gave rise to the mammals. However, the Karoo is a vast thickness of rock which spans the Permian and Triassic periods, so cannot really be described as a single Fossil-Lagerstätte. In this chapter the famous Lagerstätte of Grès à Voltzia, in the northern Vosges mountains of north-eastern France (**118**), is described. As mentioned in Chapter 6, there are certain similarities with the Late Carboniferous Mazon Creek Lagerstätte (the Grès à Voltzia was laid down in a deltaic environment, for example), but there are preservational differences and, of course, the living biota of the Triassic was different from that of the Carboniferous.

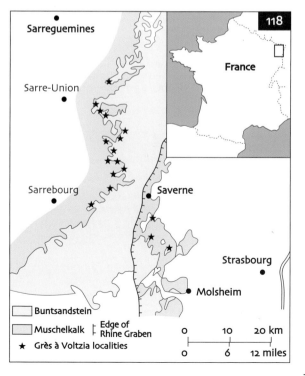

118 Map showing fossil localities in the Grès à Voltzia Sandstone in the northern Vosges Mountains, northwest of Strasbourg, France (after Gall, 1985).

History of discovery of the Grès à Voltzia biota

Triassic sandstone has been quarried in the northern Vosges for centuries, for building (e.g. the magnificent Gothic cathedral in Strasbourg, **119**) and for millstones. Associated with the sandstones (the Grès à meules or millstone grit) are clayrocks which are a nuisance to the quarrymen, but are rich in fossils. In Britain, Triassic sandstones are also extensively quarried as building stones, but few make good freestones (stone which lack a prominent 'grain' and so can be carved in any direction into intricate decorative tracery). Moreover, the clay wayboards (lenses of clay and silt intercalated in the sandstones) in the British Triassic are strongly oxidized so preserve fossils poorly, while those in the Vosges are not oxidized and special conditions prevailed which allowed fine preservation. Systematic collection of the fossils began in the middle of the twentieth century by Louis Grauvogel of the Louis Pasteur University, Strasbourg. Léa Grauvogel-Stamm, Louis's daughter, joined in the research with her interest in palaeobotany, and in 1971 Jean-Claude Gall became another member of the research team on the Grès à Voltzia biota and its palaeoecology, which continues today.

Stratigraphic setting and taphonomy of the Grès à Voltzia biota

The Triassic Period was named by von Alberti (1834) because of its three-fold division in Central Europe into a lower Buntsandstein (sandstone), a middle Muschelkalk ('mussel-chalk' – a marine limestone), and an upper Keuper Sandstein (sandstone). In Britain the Muschelkalk is missing but the name Triassic has been accepted into international stratigraphical usage. The Grès à Voltzia is in the upper part of the Buntsandstein. The lower unit of the Grès à Voltzia, the Grès à meules, is a fine sandstone, while the upper unit, the Grès argileux, is silty and marks the beginning of the marine transgression as the Muschelkalk sea spread across the continent. In this chapter, we are concerned only with the Grès à meules, in which three facies have been recognized (Gall, 1971, 1983, 1985; **120**): (a) thick

119 The magnificent Gothic Cathedral of Notre-Dame in Strasbourg, Alsace, France, built from Grès à Voltzia sandstone from the northern Vosges mountains.

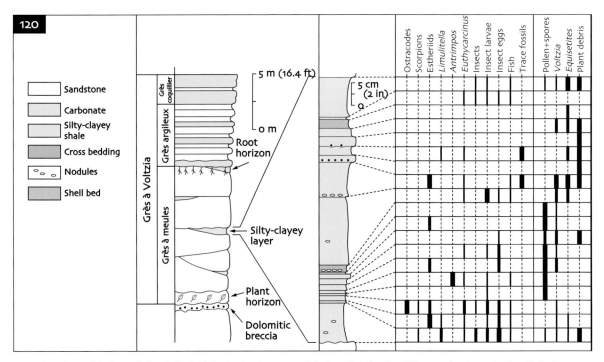

120 Stratigraphic log of the Grès à Voltzia Sandstone, with details of a fossiliferous horizon in the Grès à meules, and relative abundances of the biota (after Gall, 1971).

lenses of fine-grained sandstone, sometimes grey or pink but most often multicoloured, containing land-plant debris and amphibian bone fragments; (b) green or red silt/clay lenses, generally composed of a succession of laminae each a few millimetres thick, with well-preserved fossils of aquatic and terrestrial organisms; (c) beds of calcareous sandstone (or carbonate breccia) with a sparse marine fauna. It is in the greenish (rarely red) silt/clay laminites (b) that the beautifully preserved biota occurs.

Evidence from the sediments and fossils points to a deltaic sedimentary environment (Gall, 1971, 1983). The sandstones represent point bars deposited in strongly sinuous channels, the clay lenses represent the settling of fine material in brackish ponds, and the calcareous sandstone results from brief incursions of sea water during storms. The palaeogeographical position, the red beds and the xeromorphic nature of the land flora together suggest a semi-arid climate, although the low-lying situation suggests that the delta itself was not too arid. The climate was probably seasonal, with the pools filling during the wet season and evaporating in the dry season. The presence of desiccation cracks, reptile footprints, salt pseudo-morphs, and land plants in life position at the top of each clay lens indicates complete drying-out of the pools. Moving upwards through each clay lens a transition from aquatic to terrestrial biota is also observable (Gall, 1983).

Drying-up of the pools led to the death of the aquatic fauna. The abundance of estheriids is significant in that these crustaceans are adapted to swift completion of their life cycle in temporary water bodies. Regular high evaporation rates of the water bodies also favoured deoxygenation, consequent mass mortality of the aquatic fauna, and the rapid proliferation of microbial films. Microbial films ('veils', Gall, 1990) shielded the carcasses from scavenging activity and created, by the production of mucus, a closed environment which inhibited the decomposition of organic material. Later deposition of a new detrital load (clay, silt) buried the microbial films and the organisms (Gall, 1990).

Sediment compaction resulted in some squashing of the fossils, but in others there is three-dimensional preservation by casting with calcium phosphate, which is a rare casting material in invertebrates. While it is present in organic tissues, when liberated it is swiftly recycled by other organisms. However, because of the exceptional taphonomic conditions present in the Grès à meules, rapid phosphatization occurred. Phosphatization requires a low-oxygen environment and abundant organic matter. The microbial film would have sealed the phosphates being released from the organic matter in the decaying animals, thus preventing its reuse by other organisms. Acidic conditions produced by decaying organic matter would have released free calcium ions, which combined with the phosphate to form apatite. Once a phosphatic nodule had formed, it would have prevented further flattening of the body during sediment compaction.

Thus, the taphonomy of the Grès à Voltzia is quite different from those of other Fossil-Lagerstätten. While there are similarities – for example, in the rapid burial,

reducing conditions and fine sediment – between Grès à Voltzia and Mazon Creek, no microbial mat or phosphatization has been reported for the latter. Phosphatization occurs in many Lagerstätten, e.g. Santana (Chapter 11), but the process is different to that seen in the Grès à Voltzia.

Description of the Grès à Voltzia biota

Plants. The Grès à Voltzia is named for the abundant remains of the conifer *Voltzia heterophylla* (**121**). This is a bushy conifer which formed thickets between the delta distributaries together with other gymnosperms such as *Albertia* (**122**), *Aethophyllum* and *Yuccites*. On the banks of the waterways were probably thick stands of vegetation which was adapted to rooting in shifting sand substrates and frequent flooding, such as horsetails

121 The conifer *Voltzia heterophylla* (GGUS). Scale bar is 10 mm (0.4 in).

122 The gymnosperm *Albertia*. (GGUS). Scale bar near base of stem is 10 mm (0.4 in).

(*Equisetum, Schizoneura*). Ferns such as *Anomopteris* (**123**) and *Neuropteridium* were also present, and cycads and ginkgos have also been reported. Many of the macroplant remains occur in the coloured sandstone lenses (facies (a), above) together with amphibian remains, although drifted plant debris also occurs in the silt-clay laminites.

Cnidarians. A medusoid, *Progonionemus vogesiacus*, is known from 10 juvenile to adult specimens (Grauvogel and Gall, 1962). It consists of a bell-shaped umbrella about 8–40 mm (0.3–1.6 in) across, and many tentacles of 9–40 mm (0.4–1.6 in) in length. Both primary and secondary tentacles can be recognized, and gonads can be seen in the adult forms. *Progonionemus* was identified as close to *Gonionemus*, belonging to the group Limnomedusae, which includes fresh- and brackish-water forms.

Brachiopods. The inarticulate brachiopod *Lingula tenuissima* occurs in the Grès à meules. *Lingula* is a burrowing brachiopod which is generally found in shallow, brackish-water settings which most articulate brachiopods cannot tolerate. Most interestingly, *Lingula* is preserved in its upright, life position, indicating that it lived exactly where it died (autochthonous) rather than being drifted in by currents.

Annelids. A few specimens of annelids have been described by Gall and Grauvogel (1967), including the polychaetes *Eunicites* and *Homaphrodite*. These marine worms also occur in waters of variable salinity.

Molluscs. A variety of bivalves and gastropods has been recorded from the Grès à meules, including 'pectens' (free-swimming forms) and the trigonioid *Myophoria*. In general, these animals are rare in the brackish-water pools, but indicate marine water nearby in time and space.

Arthropods. These are the most abundant animals found in the laminites and, with their brown cuticles (probably not the original coloration), their remains look remarkably fresh. The commonest arthropods are those associated with water sometime in their life cycle. The horseshoe crab (limulid) *Limulitella bronni* (**124**) is a common component of the fauna of the Grès à meules; its trackways (*Kouphichnium*) occur as well as remains of the dead animals and moulted exoskeletons. Horseshoe crabs are predominantly marine, but can be found a great distance up rivers, especially when they come ashore in hordes to mate, and in late Carboniferous times (e.g. Mazon Creek, Chapter 6) they appear to have tolerated both fresh and saline water, and were possibly amphibious.

123 The fern *Anomopteris* (GGUS). Frond is about 140 mm (5.5 in) across.

124 The horseshoe crab *Limulitella bronni* (GGUS). Length of animal (including tail spine) is 55 mm (2.2 in).

Horseshoe crabs are the only primarily aquatic members of the arthropod subphylum Chelicerata that are still alive today. In the middle of the Palaeozoic Era another chelicerate group, the scorpions, were also living in water, but by Carboniferous times they had left the water and become terrestrial. Scorpions occur in the Grès à meules (**125**) and form part of the terrestrial fauna. They have been placed in the ancient fossil family Eoscorpiidae. Another, perhaps more familiar, group of Chelicerata is the Araneae: spiders (**126**). About a dozen specimens are known from the Grès à meules, and these were described by Selden and Gall (1992) as the oldest representative of the mygalomorph spiders (trapdoor, funnel-web and tarantula spiders). The single species, *Rosamygale grauvogeli*, was placed in the modern family Hexathelidae, a mainly Gondwanan family today, although some species occur in the Mediterranean region.

Among Crustacea the notostracan *Triops cancriformis* occurs in the Grès à meules. Today, these animals are characteristic of ephemeral pools, and so their presence in the temporary water bodies of the Grès à Voltzia environment is to be expected. Conchostracans (estheriids) are also common, and these small, bivalved crustaceans are frequently found in lacustrine and other non-marine environments. Other crustaceans found in the Grès à meules are the mysid shrimp *Schimperella*, the isopod *Palaega pumila* (Gall and Grauvogel, 1971), and some crayfish: *Antrimpos* (**127**), a predominantly nektonic form, and the benthic *Clytiopsis*. These crustaceans are a common component of the fauna of the temporary pools, and many were originally described by Bill in 1914.

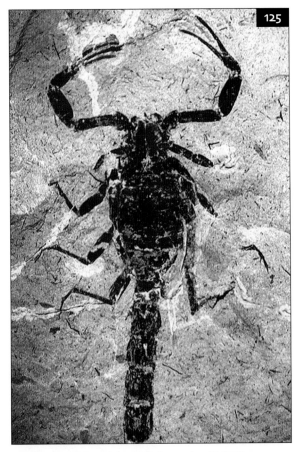

125 Scorpion from the Grès à meules (GGUS). Length of body is 60 mm (2.4 in).

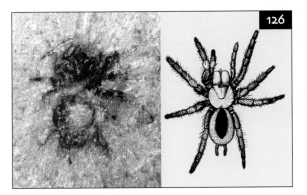

126 The funnel-web spider *Rosamygale grauvogeli*. Photograph (left) and reconstruction (right) (GGUS). Length about 6 mm (0.2 in) (after Selden and Gall, 1992).

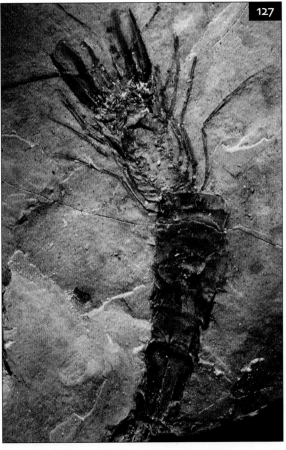

127 The crayfish *Antrimpos* (GGUS). Length (excluding antennae) 50 mm (2 in).

A group of rather strange crustaceans, now extinct, is the Cycloidea. These range from the early Carboniferous to the Cretaceous, and are found world-wide. The genus *Halicyne* (**128**) occurs in the Grès à meules. Cycloids resemble modern fish-lice and, indeed, are thought by some palaeontologists to be external parasites of fish. They are characterized by a nearly circular carapace with only small legs protruding from beneath the rim.

Myriapods are represented by some rather beautifully preserved specimens of millipede (**129**), which have yet to be formally described. Millipedes are important feeders on detrital plant material at the present day.

A group of arthropods which ranged from possibly as old as the Cambrian (as evidenced by possible trackways) to the Triassic Period and which, though crustacean-like in appearance, have some affinities with insects, is the Euthycarcinoidea (see **94**). These animals have a head with antennae, a long body with many legs, a short 'abdomen' and a spine-like tail. A few specimens of *Euthycarcinus* are known from the Grès à meules. These were the first euthycarcinoids to be described, and are among the youngest known examples of the group (Wilson and Almond, 2001).

Insects in the Grès à meules include both aquatic and terrestrial forms. Many terrestrial insects have aquatic larvae, and these are well represented in the beds (**130**, **131**). Their presence indicates fresh, rather than brackish, water. The insects present include: Ephemeroptera (mayflies), Odonata (dragonflies, including a large species), Blattodea (cockroaches), Coleoptera (beetles), Mecoptera (scorpion flies), Diptera (true flies), and Hemiptera (bugs).

In addition to the adults and larvae of insects and crustaceans, masses of eggs of these arthropods have been found in the Grès à meules. Gall and Grauvogel (1966) described a variety of different types of eggs (*Monilipartus*, *Clavapartus*, *Furcapartus*) which seem most similar to the eggs of chironomid midges. These eggs look like small, dark, circular or oval spots and occur in chains (*Monilipartus*) or discrete packages (*Clavapartus*, *Furcapartus*) as if they were enclosed in mucus in life. Some eggs also occur within or associated with the shells of conchostracans and so could represent the eggs of these animals.

Fish. A few fish occur in the Grès à meules; juveniles are especially abundant. Actinopterygians are represented by a single specimen of *Saurichthys*, many specimens (mainly juveniles) of *Dipteronotus* (**132**) and a few dozen examples of the holostean *Pericentrophorus*. Coelacanth scales also occur. All of these fish are nektonic (i.e. swimming rather than bottom-living) forms. The shark egg-case *Palaeoxyris*, which also occurs at Mazon Creek (Chapter 6), is common in the Grès à meules, as are clutches of eggs of other fish.

Tetrapods. Amphibian remains are quite common in the sandstone facies (a). They have been attributed to the capitosauroid temnospondyls *Odontosaurus* and *Eocyclotosaurus*. These were aquatic animals with flattened, triangular heads, and the group was entirely Triassic. Reptile remains are rare in the Grès à Voltzia, but *Chirotherium* trackways, which are common in the Triassic of Britain, provide evidence for the presence of fairly large animals.

Palaeoecology of the Grès à Voltzia biota

Much of the biota of the Grès à Voltzia has yet to be described systematically but, in contrast, the palaeoecology of the Lagerstätte has been well studied and the whole scenario is well worked out, particularly by many years of research by Jean-Claude Gall of the University of Strasbourg (Gall, 1971). The interdigitation of the three facies in the lower Grès à Voltzia indicates deposition in a complex deltaic setting adjacent to the sea. Following the facies changes upwards, we witness the gradual change from an alluvial plain to a delta, and finally a marine transgression by the Muschelkalk Sea.

Facies (a) of the Grès à meules represents fluvial distributary channels on the delta top. Lenses a few metres thick of fine-grained pink or grey sandstone with an erosive base represent channel bars, and coarser-grained, poorly sorted sandstones with plant and amphibian debris and mud-flake conglomerates are interpreted as crevasse splay deposits. Proximal sandstones contain coarser bone and vegetation debris than do distal.

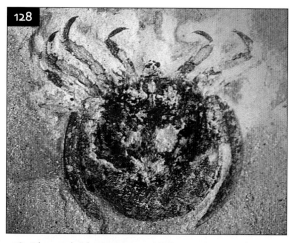

128 The cycloid crustacean *Halicyne ornata* (GGUS). Width of body 15 mm (0.6 in).

129 A juliform millipede (GGUS). Length about 50 mm (2 in).

Facies (b) provides evidence of temporary pools. Lenses of green or red silty clay occur as intercalations within the sandstones. They consist of a series of graded laminations, each layer usually only a few millimetres thick. The sediment was deposited by flood water spilling over from river channels or as a result of exceptionally high tides. The occurrence of salt pseudomorphs and mud cracks is evidence for desiccation (i.e. periodic emergence); increased salinity is reflected in the high boron content of the clay minerals (Gall, 1985). Individual pools probably existed for only a short period of time – a few weeks to several seasons. One 60 cm (2 ft) lens preserves what appears to be a single annual cycle in gymnosperms, from inflorescence to seed (Gall, 1971). The tops of the fine-grained lenses usually show signs of emergence, such as mud cracks and plant-root traces (Gall, 1971, 1985). Aquatic organisms are concentrated near the bases of the lenses, which presumably reflects periods when the water was deeper and more permanent. Dramatic changes in the proportions of different faunal elements from lamina to lamina within the lenses (Gall, 1971) indicate that a new assemblage was established with each initial influx of water (and accompanying sediment).

It is clear that the aquatic fauna is autochthonous and that the terrestrial biota was derived from the immediate vicinity of the stagnating pools. The evidence includes the preservation of organisms in life position (e.g. *Lingula*), the absence of any current directional orientation, the presence of both larval and adult forms in the same assemblage, and the concentration of organisms in former residual ponds. The fauna, which was dominated by euryhaline forms (those tolerant of a wide range of salinities), was subsequently killed (sometimes with clear evidence of mass mortality), probably by shifts in oxygen levels associated with a decrease of the size of the water body through evaporation.

Facies (c), calcareous sandstones or brecciated carbonate, represents temporary marine incursions. The carbonate breccias are storm deposits and the rarer lenses of calcareous sandstone containing gastropods suggest occasional, minor marine transgressions.

The palaeoecological scenario which can be envisaged for the Grès à Voltzia delta is one of low-lying sandy substrates, with thickets of gymnosperms (e.g. *Voltzia, Yuccites*) and lycopods (e.g. *Pleuromeia*), with horsetails and ferns (e.g. *Neuropteridium, Anomopteris*) between. The vegetation appears to have been restricted to just a few main species: those which were able to root in relatively unstable sand and could survive (or quickly recolonize after) flooding. Amongst this vegetation there was a similarly species-poor fauna of insects (mostly those with aquatic larvae), millipedes, scorpions, spiders (*Rosamygale*), amphibians and reptiles. In the brackish-water pools there was an abundance of life, but each taxon exhibited low species diversity and many animals were euryhaline forms. The fauna included medusae (*Progonionemus*), polychaete annelids, *Lingula*, some bivalves, *Limulitella*, crustaceans (e.g. *Triops*, conchostracans, *Schimperella, Antrimpos, Clytiopsis, Halicyne*), euthycarcinoids, fish (e.g. *Dipteronotus*), and insect eggs (e.g. *Monilipartus*) and larvae. Gall (1985, figs. 8, 9) illustrated reconstructions of the palaeoecology of the Grès à meules.

Low diversity is characteristic of both semi-arid terrestrial and brackish-water communities, and it would appear that, thanks to the exceptional preservation of soft-bodied organisms, there is little biota absent from the reconstruction. Moreover, the low diversity, sandy substrate, semi-arid (seasonal) ecosystem was well established by the Triassic Period and continues, albeit with somewhat different taxic composition, today.

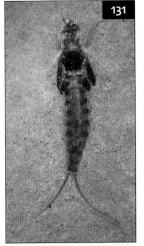

130 Insect larva (GGUS). Body length about 17 mm (0.7 in).

131 Insect larva (GGUS). Length (including tails) about 15 mm (0.6 in).

132 Juvenile fish *Dipteronotus* (GGUS). Length 35 mm (1.4 in).

Comparison of Grès à Voltzia with other biotas

Briggs and Gall (1990) made a quantitative comparison of the Grès à Voltzia biota with four major Carboniferous Lagerstätten, to test and extend Schram's (1979) concept of a faunal continuum among these Carboniferous Lagerstätten. A particularly important aspect of this work was the realization that not only should comparison be made between the taxic composition of the biotas but also the stratigraphic position (especially when comparing biotas either side of an extinction event), palaeo-environmental differences, and preservational effects should be evaluated. Using a new coefficient of similarity, these authors compared the Grès à Voltzia with the Namurian Bear Gulch Member of the Heath Formation of Montana, the Stephanian Montceau-les- Mines Lagerstätte of France, the Lower Carboniferous Glencartholm Volcanic Beds of Scotland, and the Westphalian Mazon Creek biota of Illinois. The results showed that the Grès à Voltzia was much closer to Mazon Creek than to the other biotas, which ranked (in decreasing similarity): Glencartholm, Bear Gulch, and Montceau-les-Mines.

Stratigraphic age clearly had little effect on the results: Glencartholm is the furthest, stratigraphically, from the Grès à Voltzia and yet came second-most similar, while the youngest biota (and closest geographically), Montceau-les-Mines, was least similar. Taphonomy was important in that with better preservation more taxa are preserved. So, for example, the taphonomy of Glencartholm is not good enough to preserve organims without some sort of cuticle, and medusae, polychaetes, and many terrestrial forms, such as spiders and insects, are absent from this Lagerstätte, even though they were probably around at the time of deposition. The factor which came out as most influential in determining similarity was palaeoenvironment. Sedimentological evidence from both Mazon Creek (Chapter 6) and Grès à Voltzia suggests palaeo-environmental settings transitional between terrestrial and a marine-influenced delta. Taxonomic groups common to both Lagerstätten include medusoids, brachiopods, polychaete annelids, bivalve and gastropod molluscs, horseshoe crabs, scorpions, spiders, ostracods, shrimps, cycloids, euthycarcinoids, millipedes, insects, fish and tetrapods. Organisms adapted to fluctuating conditions (e.g. euryhaline species, plants adapted to unstable substrates) show strong congruence at the family and lower levels between the Lagerstätten, and these groups were little affected by the end-Permian extinction event. The main taxonomic differences between the Grès à Voltzia and Mazon Creek lie in the higher crustaceans and the insects. Many of those represented in the Grès à Voltzia appeared in the Permian and radiated across the Permo–Triassic boundary as Palaeozoic forms became extinct. Thus, Briggs and Gall (1990) concluded that there is a striking continuity between the biotas of Carboniferous and Triassic transitional sedimentary environments.

Further Reading

Bill, P. C. 1914. Über Crustaceen aus dem Voltziensandstein des Elsasses. *Mitteilungen der Geologisches Landesanstalt von Elsass-Lothringen* **8**, 289–338.

Briggs, D. E. G. and Gall, J.-C. 1990. The continuum in soft-bodied biotas from transitional environments: a quantitative comparison of Triassic and Carboniferous Konservat-Lagerstätten. *Paleobiology* **16**, 204–218.

Gall, J.-C. 1971. Faunes et paysages du Grès à Voltzia du nord des Vosges. Essai paléoécologique sur le Buntsandstein supérieur. *Mémoires du Service de la Carte Géologique d'Alsace et de Lorraine* **34**, 1–318.

Gall, J.-C. 1972. Fossil-Lagerstätten aus dem Buntsandstein der Vogesen (Frankreich) und ihre ökologische Deutung. *Neues Jahrbuch für Geologie und Paläontologie, Monatshefte* **1972**, 285–293.

Gall, J.-C. 1983. The Grès à Voltzia delta. 134–148. *In* Gall, J.-C. *Ancient sedimentary environments and the habitats of living organisms.* Springer-Verlag, Berlin, xxii + 219 pp.

Gall, J.-C. 1985. Fluvial depositional environment evolving into deltaic setting with marine influences in the Buntsandstein of northern Vosges. 449–477. *In* Mader, D. (ed.). *Aspects of fluvial sedimentation in the Lower Triassic Buntsandstein of Europe.* Lecture Notes in Earth Sciences 4. Springer-Verlag, Berlin, viii + 626pp.

Gall, J.-C. 1990. Les voiles microbiens. Leur contribution à la fossilisation des organismes au corps mou. *Lethaia* **23**, 21–28.

Gall, J.-C. and Grauvogel, L. 1966. Ponts d'invertébrés du Buntsandstein supérieur. *Annales de Paléontologie (Invertébrés)* **52**, 155–161.

Gall, J.-C. and Grauvogel, L. 1967. Faune du Buntsandstein. III. Quelques annélides du Grès à Voltzia des Vosges. *Annales de Paléontologie (Invertébrés)* **53**, 105–110.

Gall, J.-C. and Grauvogel, L. 1971. Faune du Buntsandstein. IV. *Palaega pumila* sp. nov., un isopode (Crustacé Eumalacostracé) du Buntsandstein des Vosges (France). *Annales de Paléontologie (Invertébrés)* **57**, 77–89.

Gall, J.-C. and Grauvogel-Stamm, L. 1984. Genèse des gisements fossilifères du Grès à Voltzia (Anisien) du nord du Vosges (France). *Géobios, Mémoire Special* **8**, 293–297.

Grauvogel, L. and Gall, J.-C. 1962. *Progonionemus vogesiacus* nov. gen, nov. sp., une méduse du Grès à Voltzia des Vosges septentrionales. *Bulletin du Service de la Carte Géologique d'Alsace et de Lorraine* **15**, 17–27.

Grauvogel-Stamm, L. 1978. *La flore du Grès à Voltzia (Buntsandstein supérieur) des Vosges du Nord (France): morphologie, anatomie, interprétations phylogénique et paléogéographique.* Mémoires des Sciences Géologiques, n. 50. Institut de Géologie de l'Université Louis Pasteur, Strasbourg.

Schram, F. R. 1979. The Mazon Creek biotas in the context of a Carboniferous faunal continuum. 159–190. *In* Nitecki, M. H. (ed.). *Mazon Creek Fossils.* Academic Press, New York, 581 pp.

Selden, P. A. and Gall, J.-C. 1992. A Triassic mygalomorph spider from the northern Vosges, France. *Palaeontology* **35**, 211–235.

Wilson, H. M. and Almond, J. E. 2001. New euthycarcinoids and an enigmatic arthropod from the British Coal Measures. *Palaeontology* **44**, 143–156.

THE HOLZMADEN SHALE

Chapter Eight

Background: the Mesozoic marine revolution

From the late Triassic onwards, while dinosaurs were dominating the land and pterosaurs were taking to the air, a revolution in marine life was taking place beneath the waves. As the supercontinent of Pangaea began to break up, sea levels rose, and vast areas of low-lying land were flooded by epicontinental seas which supported coral reefs teeming with life. In such environments, full of potential prey, marine reptiles quickly diversified and dominated the seas; they included ichthyosaurs, plesiosaurs, crocodiles and turtles.

The ichthyosaurs and plesiosaurs are not closely related, but both probably belong to a group of diapsid reptiles called the lepidosaurs, which today includes the lizards and snakes. The crocodiles are diapsid archosaurs (along with pterosaurs and dinosaurs), while the turtles belong to the more primitive anapsid reptiles.

Ichthyosaurs were fish-shaped reptiles (**138**), entirely adapted to life in the sea, with streamlined bodies and paddle-like limbs, looking very much like modern sharks (which are fish) or dolphins (which are mammals). They had long, thin snouts lined with sharp, conical teeth and fed on fish and cephalopods. Their large eye sockets suggest that they had good eyesight – perhaps needed in the muddy waters of the Jurassic sea – and a ring of bony plates surrounding the eye (the sclerotic ring) may have been used to change the focal length, rather like a zoom lens. They swam by beating the flexible body and powerful tail from side to side, like a shark, and they used the small front paddles for steering and a dorsal fin for balance. They were so well adapted to life in the sea that unlike most marine reptiles they could not come ashore to lay eggs, but gave birth to live young (like whales and most sharks).

The plesiosaurs were the true giants of the Mesozoic seas. Their bodies were broad and flat (**141**) and they swam by rowing the body along by means of two pairs of paddle-like limbs, or alternatively by subaqueous flight, in the manner of turtles and penguins. The tail served merely as a rudder.

There are two groups; the true plesiosaurs had long, flexible necks (up to 72 cervical vertebrae) and very small heads. They fed mainly on small fish in inshore waters, using the long neck like a snake to dart after fast-moving prey. The long, slim, pointed teeth were homodontous (all the same type). A second group, the pliosaurs, had larger heads (up to 4 m [13 ft] long) and shorter necks (as few as 13 cervical vertebrae). But the real difference was in their dentition, which was heterodontous (with different types of teeth). The giant pliosaurs, up to 13 m (43 ft) long, were the top predators of the Jurassic seas and preyed on other marine reptiles in the open sea, just as killer whales today feed on smaller whales and seals. With their more powerful paddles it is possible that the plesiosaurs could have struggled ashore to lay eggs.

The only invasion of the sea by the archosaurs was the development in the Jurassic of the short-lived group of marine crocodiles. Crocodiles and alligators are arguably the most successful four-footed hunters that have ever lived on Earth; they can equally well walk and swim and catch prey on land and in water. Most of the crocodiles of the Jurassic belong to the semi-aquatic teleosaurs, which were similar to the gavials of the River Ganges in India today. Evolving from the teleosaurs, however, was a highly specialized marine group, the metriorhynchids, which flourished in the later Jurassic and persisted into the early Cretaceous. They are so distinctive in their marine adaptations that some authorities class them as a separate group, the thalattosuchians, or sea crocodiles. They were unarmoured, their limbs were modified from walking legs into swimming paddles, and the tail developed a fish-like fin, akin to that of the ichthyosaurs. The metriorhynchids were open-sea hunters, and stomach contents include indigestible pterosaur bones and hooks from belemnoid or squid tentacles.

In the Jurassic seas dominated by ichthyosaurs, plesiosaurs and crocodiles were also turtles and a variety of fish which were going through their own revolution. They included actinopterygians (ray-finned fish), sarcopterygians (lobe-finned fish) and chondrichthyans (cartilaginous fish) (see Chapter 4). While most of the early Jurassic actinopterygians were of the more primitive groups, either holosteans (i.e. the bony ganoid fish) or chondrosteans (i.e. the sturgeons), a new group of advanced fish, the teleosteans (or modern bony fish), became dominant by the end of the period.

Lower Jurassic marine ecosystems are known from many locations across Europe, but around the small village of Holzmaden in the Schwäbische Alb of Baden-Württemberg, southern Germany (**133**), the Posidonienschiefer (*Posidonia* Shale) contains an abundant and sometimes completely preserved biota within black bituminous marls. All the major groups of marine reptiles and fish are exquisitely preserved, often with the outline of their skin and soft body tissue clearly visible. In addition there are rare pterosaurs and dinosaurs and a variety of marine invertebrates, dominated by coleoid cephalopods such as squids and belemnoids, sometimes complete with ink sacs and tentacles.

History of discovery and exploitation of the Holzmaden Shale

Shale has been quarried around the villages of Holzmaden, Ohmden, Zell and Boll to the south-east of Stuttgart since the end of the sixteenth century. The shale, known as 'Fleins', was initially used for roofing and paving, but because of its poor resistance to weathering it was later confined to internal use, such as oven bases, hearths, window sills, wall cladding, flooring, wash stones, tanner's slates and laboratory tables. At Holzmaden the Fleins is a regular, 18 cm (7 in) thick bed which splits into four equal layers. Within the same sequence limestone was also quarried as a building stone for cellars.

The shale itself is bituminous and contains up to 15% organic matter. This has resulted in the past in several serious shale fires when excavations had been neglected. In 1668 a shale digging at Boll burned for six years during which time oil flowed from the burnt shale and was sold locally. The last big shale fire was at Holzmaden from 1937 to 1939.

During times of emergency and especially during the First World War, this shale oil has been utilized as an alternative energy source by companies such as the Jura Oil Shale Works at Göppingen; the shale yields a maximum of 8%. After the war production ceased, but the bituminous shale continued to be used for a short time as a heat source for the manufacture of cement from the White Jura Limestone and Liassic Marl. Just prior to the Second World War, the Portland cement works in the Balingen area produced shale oil as a by-product, and during the war a crude oil plant was again established for the production of shale oil. Oil and tar extracted from the *Posidonia* Shale are now used in the pharmaceutical industry, and at Bad Boll Spa the shale is finely ground and marketed as a medicinal mud!

The manual quarrying of former years brought to light many remarkable fossils from the Holzmaden Shale. In 1939 about 30 quarries employed 100 workmen, but unfortunately this came almost to a complete stop during the Second World War. In 1950 about 20 quarries reopened, but since then the industry has suffered from competition from imported slate, marble and synthetic material; nowadays only a handful of quarries remain (**134**) and because their operations are largely mechanized, fewer fossils are found. The area is now protected by law, but some quarries are open to collectors.

Fossils were first discovered in this area during excavations at the spa town of Boll in 1595, but it was a

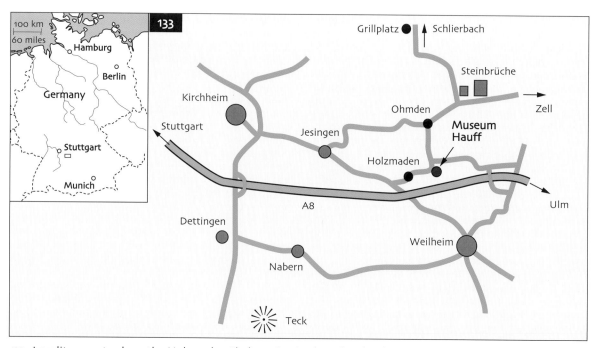

133 Locality map to show the Holzmaden Shale region in the Schwäbische Alb of southern Germany.

134 Quarry in Holzmaden Shale (Schieferbruch Kromer) at Ohmden in the Schwäbische Alb of southern Germany.

135 Laminated bituminous marls with intercalated limestones at Schieferbruch Kromer, Ohmden.

sensational discovery in 1892 which brought them to the attention of the wider scientific world. Bernhard Hauff (1866–1950), the son of a chemist who had come to Holzmaden in 1862 to pursue his interest in extracting oil from shale, found many new fossils in his father's quarry which he prepared meticulously. In 1892 he succeeded in exposing the outline of the body of a fossil ichthyosaur complete with its skin. Previous reconstructions of ichthyosaurs had shown them without dorsal fins and with a long, thin tail, but Hauff's specimen proved the presence of a triangular dorsal fin and a fleshy upper lobe to the tail, neither of which had skeletal support.

Bernhard Hauff and his own son, Bernhard Hauff junior (1912–1990), built the Urwelt-Museum Hauff at Holzmaden to exhibit the finest fossils from this unusual Fossil-Lagerstätte (see Hauff and Hauff, 1981).

Stratigraphic setting and taphonomy of the Holzmaden Shale

The Holzmaden Shale is the informal name given to the Posidonienschiefer at its outcrop in the Holzmaden area. Here it consists of 6–8 m (20–26 ft) of black, bituminous marls (135) with intercalated limestones of Lower Toarcian age (Lower Jurassic; zones of *Dactylioceras tenuicostatum, Harpoceras falcifer* and *Hildoceras bifrons*; approximately 180 million years ago). The Tübingen geologist August Quenstedt subdivided the Jurassic of the Schwäbische Alb into six divisions termed alpha to zeta, and the Holzmaden Shale falls within the 'Lias epsilon (ε)'.

It has been further subdivided (see Hauff and Hauff, 1981, p. 10) into the lower (εI), middle (εII) and upper (εIII) divisions (136). The Fleins (see p. 80) occurs in the middle epsilon (εII 3). Above the Fleins is the 'Untere Schiefer' (εII 4), from which the shale oil was extracted, and it is from the lower part of this layer that the best preserved fossils occur, including the ichthyosaurs with soft tissue preservation of skin and muscles. Above this is the 'Untere Stein' (εII 5), a hard and weather-resistant limestone formerly used as a building stone (p. 80), which has yielded uncompressed fossil fish. The 'Oberer Stein' (εII 8) is another

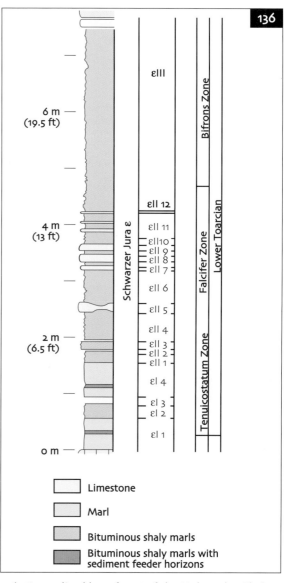

136 Generalized log of part of the Holzmaden Shale sequence in the Holzmaden area (after Wild, 1990).

Legend for figure 136:
- Limestone
- Marl
- Bituminous shaly marls
- Bituminous shaly marls with sediment feeder horizons

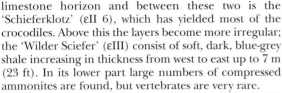

limestone horizon and between these two is the 'Schieferklotz' (εII 6), which has yielded most of the crocodiles. Above this the layers become more irregular; the 'Wilder Sciefer' (εIII) consist of soft, dark, blue-grey shale increasing in thickness from west to east up to 7 m (23 ft). In its lower part large numbers of compressed ammonites are found, but vertebrates are very rare.

At the beginning of the Jurassic the Schwäbische Alb lay under a wide epicontinental sea which was eventually to gain connections with the Tethys Ocean to the south (see also p. 101). The extensive sea, which covered much of northern Europe, was divided into a number of basins separated by submarine highs and islands; the Posidonienschiefer in the vicinity of Holzmaden was deposited in the South German Basin between Ardennes Island to the west and the Vindelicisch Land/Bohemian High to the south and east (Hauff and Hauff, 1981, fig. 4).

The dark colour of the fine-grained marls is caused partly by diffused pyrite and partly by the high percentage (up to 15%) of solid organic matter (kerogen and poly-bitumen). Together these suggest deposition within a stagnant basin, starved of oxygen and rich in H_2S. As with the Burgess Shale (Chapter 2) and the Solnhofen Limestone (Chapter 10) the marls are finely laminated and individual laminae can be traced over large distances, also suggesting deposition in still waters. Generally there is no evidence of bioturbation of the sediment, and benthic organisms are extremely rare, being restricted to a few echinoids, crustaceans and burrowing bivalves.

Bottom conditions were clearly hostile to life, the absence of bottom currents leading to an anoxic seabed with insufficient oxygen to support life – other than anaerobic bacteria – or to enable aerobic bacterial decay of dead organic tissue. Fresh water flowed into the South German Basin in the upper water zones only, and these well-oxygenated and nutritious surface waters supported a rich nektonic (swimming) and planktonic (floating) fauna. Occasional bioturbated horizons (εI 3, εI 4, top of εIII) do suggest temporary periods favourable to epibenthic life and alignment of fossils at certain horizons similarly suggests occasional bottom currents, but generally there was no exchange of water near the seabed.

Micro-organisms falling to the seabed would have begun to decay, but this would soon have used up the available oxygen so that organic particles were instead incorporated into the sediment. Carcasses of macro-organisms, such as marine reptiles and fish, were thus buried in anoxic mud and decay of their soft tissue was arrested. Scavengers would also have been excluded by the noxious environment so that the carcasses were preserved intact. Such a stagnation depositional model for the Holzmaden Shale compares with the conditions today in the Black Sea, an epicontinental basin with a restricted aperture.

Kauffman (1979) contested this model and suggested that some of the presumed pseudoplanktonic fauna (especially bivalves and crinoids, see p. 86) was in fact benthic and that benthic communities characterized much of the depositional history of the shale. He suggested that only the sediment was anoxic and that the water body just above the sediment/water interface was hospitable. This has not been generally accepted, but it did lead to a modification of the stagnation model (Brenner and Seilacher, 1979; Seilacher, 1982) in which stagnant conditions were sometimes interrupted by high-energy events related to severe storms which had brief oxygenating effects.

One aspect of Kauffman's model that is interesting is his proposal of the development of an algal mat just above the sediment/water interface, which if nothing else would inhibit grazers and scavengers. The presence of such a cyanobacterial mat has also been suggested in the case of Ediacara (Chapter 1), Grès à Voltzia (Chapter 7), the Solnhofen Limestone (Chapter 10) and the

137 The ichthyosaur *Stenopterygius quadriscissus*, with the body outline preserved as a black organic film (UMH). Length 120 cm (4 ft).

138 Reconstruction of *Stenopterygius*.

Santana/Crato formations (Chapter 11) and is an important contributory factor in the preservation of soft tissue.

Description of the Holzmaden Shale biota

Ichthyosaurs. The Holzmaden Shale is famous for its complete vertebrate skeletons which are often preserved with the skin of the animals marking the outline of the body as a black film. This is seen in sharks, in bony fish, in crocodiles and pterosaurs, but most spectacularly in the various species of the ichthyosaur *Stenopterygius* (**137**). Bernhard Hauff's initial discovery of this phenomenon (p. 81) proved that ichthyosaurs had a dorsal fin and that their tail had an upper lobe (**138**),

neither of which have skeletal support and are therefore normally not fossilized. Many individuals of *Stenopterygius* are preserved with stomach contents (such as the indigestible hooks of cephalopods and the thick ganoid scales of fish), and more remarkable are the numerous females preserved either in the process of giving birth or with embryos within their bodies (**139**). The high percentage of pregnant females and juveniles suggests that this area was a spawning ground to which animals periodically migrated. More than 350 specimens were recorded by Hauff (1921).

Plesiosaurs and crocodiles. These are much more rare; only 13 complete skeletons and scattered remains of plesiosaurs have ever been found. They include the long-necked plesiosaur *Plesiosaurus* (**140, 141**), and the shorter-necked pliosaurs, *Peloneustes* and

139 The ichthyosaur *Stenopterygius crassicostatus*, with five embryos preserved inside an adult, and one juvenile (UMH). Length 300 cm (10 ft).

140 The plesiosaur *Plesiosaurus brachypterygius* (UMH). Length 280 cm (9 ft).

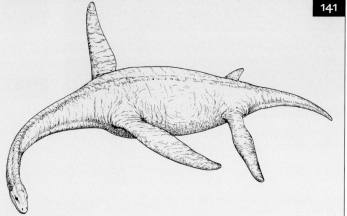

141 Reconstruction of *Plesiosaurus*.

Rhomaleosaurus. Teleosaur crocodiles are somewhat more numerous – Hauff (1921) recorded 70 specimens – the most common being *Steneosaurus* (**142**, **143**), which had a long, narrow snout with many teeth and probably caught fish by quickly snapping shut its narrow jaws. Its eyes point upwards and outwards so it probably dived under shoals of prey, coming up to attack. The powerful legs and tail suggest that it could walk on land and was a strong swimmer. The much smaller *Pelagosaurus* had its eyes arranged laterally on the cranium and was a more skilful swimmer. The more heavily armoured *Platysuchus* is much rarer, being known from only four specimens worldwide.

Pterosaurs and dinosaurs. Two genera of pterosaurs, *Dorygnathus* (**144**) and *Campylognathoides*, with wingspans of 1 m (3 ft) and 1.75 m (5 ft 9 in) respectively, are known from the Holzmaden Shale, both from complete skeletons – Hauff (1921) recorded about 10 specimens in total. The only dinosaur to be recorded is the cetiosaurid *Ohmdenosaurus*, a 4 m (13 ft) long sauropod, known from a single tibia, and named after the local village of Ohmden.

Fish. Primitive holostean fish (ganoids) are represented by well-known Liassic genera such as *Lepidotes* (a heavily-built fish up to 1 m (3 ft) in length with thick, shiny scales; **145**, **146**), *Dapedium* (a gar, with a

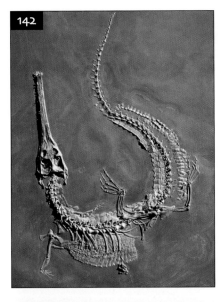

143 Reconstruction of *Steneosaurus.*

142 The teleosaur crocodile *Steneosaurus bollensis* (UMH). Length 270 cm (9 ft).

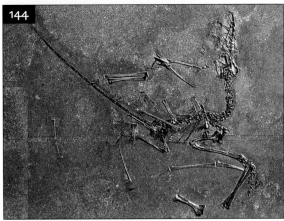

145 The holostean fish *Lepidotes elvensis* (UMH). Length 60 cm (24 in).

144 The pterosaur *Dorygnathus banthensis* (UMH). Height of block 42 cm (16.5 in).

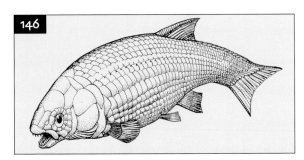

146 Reconstruction of *Lepidotes.*

flat, round body and peg-like crushing teeth; **147**, **148**), and the giant predator *Caturus*, while the rarer teleosts include the sprat-like *Leptolepis*. Occasional sharks, such as *Hybodus* (**149**) and *Palaeospinax*, are often preserved with the black outline of their skin and reached lengths of 2.5 m (8 ft), but the largest fish, up to 3 m (10 ft), were the sturgeons such as *Chondrosteus*. A single specimen of coelacanth, *Trachymetopon*, is also known.

Cephalopods. Many of the well-known Liassic ammonites, genera such as *Harpoceras* (**150**), *Hildoceras* and *Dactylioceras*, are common constituents of the Holzmaden fauna, but the most spectacular cephalopods are squids (e.g. *Phragmoteuthis*) and belemnoids (e.g. *Passaloteuthis*), often with soft body tissues preserved including ink sacs, tentacles and their hooks (**151**).

147 The holostean fish *Dapedium punctatum* (UMH). Length 33.5 cm (13 in).

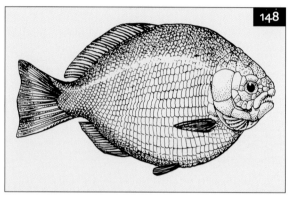

148 Reconstruction of *Dapedium*.

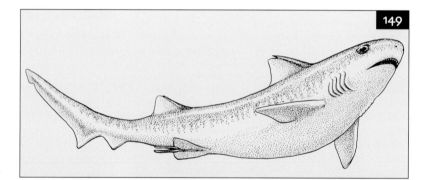

149 Reconstruction of *Hybodus*.

150 The ammonite *Harpoceras falcifer* (UMH). Diameter 20 cm (8 in).

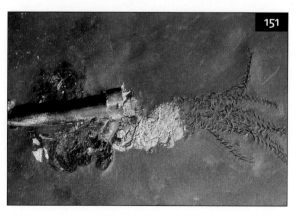

151 The belemnoid *Passaloteuthis paxillosa*, showing preservation of tentacles with their hooks (UMH). Length 23 cm (9 in).

Crinoids. The crinoids, *Seirocrinus* and *Pentacrinus*, which lived in colonies attached to floating logs (**152**), are common members of the Holzmaden fauna; a spectacular specimen of *Seirocrinus* displayed in the Urwelt–Museum Hauff in Holzmaden is over 18 m (60 ft) long, attached to a 12 m (40 ft) log.

Bivalves. Most of the common Liassic bivalves, such as *Gervillia*, *Oxytoma*, *Exogyra* and *Liostrea*, are also often found fixed by their byssal threads to floating logs or to the shells of ammonites and only occasionally to a temporarily hardened seabed. Some, such as *Bositra* (previously known as *Posidonia*), were nektoplanktonic and a few, such as *Goniomya*, were burrowers.

Plants. The Holzmaden flora comprises horsetails and gymnosperms, the latter including ginkgos, conifers and cycads.

Trace fossils. Bioturbated horizons (εI 3, εI 4, top of εIII) consist mainly of the trace fossils *Chondrites* and *Fucoides*.

The fauna and flora from the Holzmaden Shale were well documented and illustrated by Hauff and Hauff (1981).

Palaeoecology of the Holzmaden Shale

The Holzmaden Shale represents an epicontinental marine basin community living at a depth that varied between 100 and 600 m (300 and 2,000 ft), depending on basinal subsidence (Hauff and Hauff, 1981), and which was situated within the subtropical zone at about 30°N.

For much of the time stagnant and anoxic bottom waters severely limited benthic life. Occasional horizons are intensely bioturbated (for example the 'Seegrasschiefer' (εI 3), which is riddled with trace fossils such as *Chondrites* and *Fucoides*) and reflect periodic improved aeration of the seabed. Generally, however, benthic infauna (living in the sediment) was restricted to a few burrowing bivalves, such as *Solemya* and *Goniomya* (Riegraf, 1977), while the vagrant benthic epifauna (living on the sediment surface) included only minute diademoid echinoids, ophiuroids, the gastropod *Coelodiscus*, and possibly the crustacean *Proeryon*. Kauffman (1979) included some bottom-feeding fish, such as *Dapedium*, in the nektobenthos.

Well-aerated surface waters, however, supported a thriving planktonic and nektonic life. Most of the crinoids, bivalves and inarticulate brachiopods had a pseudoplanktonic lifestyle attached either to floating logs or, in the case of the latter two groups, to the shells of living ammonites (Seilacher, 1982). Kauffman (1979), however, disputed this, suggesting that the encrustations occurred on dead ammonite shells and sunken logs that were lying on the seabed and maintaining that these three groups were benthic in habit.

The bivalve *Bositra* (formerly *Posidonia*) has been interpreted as a nektoplanktonic (passively swimming) bivalve. True nektonic swimmers include the numerous ammonites and smaller fish, while the strongest swimmers were squids, belemnoids, large fish such as the sharks and sturgeons and, of course, the marine reptiles. Of these, the plesiosaurs and ichthyosaurs were inhabitants of the open sea, while the crocodiles would have lived near the coast.

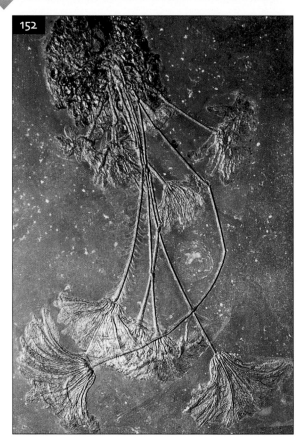

152 The crinoid *Pentacrinus subangularis*, attached to a floating log (UMH). Height 170 cm (5 ft 6 in).

The Vindelicisch landmass, some 100 km (60 miles) to the south, supported an abundance of vegetation including horsetails and a variety of gymnosperms, some of which grew into large trees (ginkgos, conifers and cycads). Leafy twigs and large logs were washed into the basin, where they were deposited as allochthonous elements of the biota along with the occasional disarticulated remains of small sauropod dinosaurs. Pterosaurs flying over the basin in search of fish were sometimes overcome by storms and drowned; their carcasses are still articulated and clearly were not transported any distance.

Trophic analysis identifies the filter feeders (crinoids and bivalves) and deposit feeders (gastropods, echinoids and ophiuroids) as the primary consumers, being preyed on by primary predators such as the bony fish and the various cephalopods. Top predators were the ichthyosaurs, plesiosaurs, crocodiles and sharks, with the large plesiosaurs at the top of the food web.

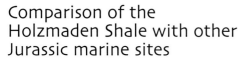

Comparison of the Holzmaden Shale with other Jurassic marine sites

Yorkshire coast, United Kingdom

The coastal outcrops of Lower Jurassic (Lias) marine sediments of the United Kingdom have been well known for many years. Those of the Dorset coast in southern England have been made famous by the celebrated nineteenth-century collector Mary Anning of Lyme Regis, who with her brother discovered one of the first ichthyosaurs and the first complete plesiosaur in rocks of Lower Lias age (Cadbury, 2000). The Liassic outcrops of the Yorkshire coast in the north-east of the country are, however, more comparable to the Holzmaden Shale since in this area it is the Upper Lias sequence that has yielded most of the marine reptiles.

Benton and Taylor (1984) listed 55 crocodiles, 69 ichthyosaurs, 33 plesiosaurs and one pterosaur from the Yorkshire sequence which, like the Holzmaden Shale, is of Lower Toarcian age (zones of *Dactylioceras tenuicostatum*, *Harpoceras falcifer* and *Hildoceras bifrons*). There are, however, significant differences in the two faunas. The majority of the specimens come from the Jet Rock and Alum Shale formations, the former consisting of hard grey bituminous shales, the latter of soft grey micaceous shales, both with bands of calcareous concretions. Typical ammonites are *Dactylioceras*, *Harpoceras*, *Hildoceras* and *Phylloceras*.

The first fossil ichthyosaur to be found in Yorkshire was a skull and partial skeleton collected in 1819, and a more complete skeleton was found in 1821. It is interesting that the original illustration of this second specimen shows the tail straightened out, as was common practice at the time, prior to Hauff's demonstration from Holzmaden specimens (p. 81) that ichthyosaurs had a bend in the vertebral column to accommodate the large tail fin. The Yorkshire ichthyosaurs belong to *Temnodontosaurus* and *Stenopterygius*.

A variety of plesiosaurs are known from Yorkshire including pliosaurs (*Rhomaleosaurus*) and true plesiosaurs (*Microcleidus* and *Sthenarosaurus*). Yorkshire crocodiles are conspecific with the Holzmaden teleosaurs; most belong to *Steneosaurus*, with *Pelagosaurus* also present, but again rare.

A single specimen of the partial skull of a pterosaur, *Parapsicephalus*, was found in 1888, and a single dinosaur bone, identified as a theropod femur, was recorded in 1926. This specimen is unfortunately lost, but if verified would be the only known record of an Upper Liassic theropod; indeed, the only other dinosaur known from the Upper Lias is that of the sauropod, *Ohmdenosaurus*, from Holzmaden (p. 84).

The Yorkshire marine reptiles are slightly younger than those from Holzmaden. Plesiosaurs and crocodiles are relatively more abundant, and ichthyosaurs are much less common in the Yorkshire Lias than at Holzmaden. Although the species of crocodile are shared between the two sites, the ichthyosaurs and plesiosaurs are generally not conspecific, with the notable exception of the ichthyosaur *Stenopterygius acutirostris*.

A large number of Yorkshire specimens are well preserved in an articulated state with little evidence of scavenging. The Jet Rock Formation, in particular, is rich in bitumen (kerogen), and Benton and Taylor (1984) suggested that bottom conditions were again anoxic.

Further Reading

Benton, M. J. and Taylor, M. A. 1984. Marine reptiles from the Upper Lias (Lower Toarcian, Lower Jurassic) of the Yorkshire coast. *Proceedings of the Yorkshire Geological Society* **44**, 399–429.

Brenner, K. and Seilacher, A. 1979. New aspects about the origin of the Toarcian *Posidonia* Shales. *Neues Jahrbuch für Geologie und Paläontologie, Abhandlungen* **157**, 11–18.

Cadbury, D. 2000. *The dinosaur hunters*. Fourth Estate, London, x + 374 pp.

Hauff, B. 1921. Untersuchung der Fossilfundstätten von Holzmaden im Posidonienschiefer des oberen Lias Württembergs. *Palaeontographica* **64**, 1–42.

Hauff, B. and Hauff, R. B. 1981. *Das Holzmadenbuch*. Repro-Druck, Fellbach, 136 pp.

Kauffman, E. G. 1979. Benthic environments and paleoecology of the Posidonienschiefer (Toarcian). *Neues Jahrbuch für Geologie und Paläontologie, Abhandlungen* **157**, 18–36.

Riegraf, W. 1977. *Goniomya rhombifera* (Goldfuss) in the *Posidonia* Shales (Lias epsilon). *Neues Jahrbuch für Geologie und Paläontologie, Monatshefte* **1977**, 446–448.

Seilacher, A. 1982. Ammonite shells as habitats in the *Posidonia* Shales of Holzmaden – floats or benthic islands? *Neues Jahrbuch für Geologie und Paläontologie, Monatshefte* **1982**, 98–114.

Wild, R. 1990. Holzmaden. 282–285. *In* Briggs, D.E.G. and Crowther, P.R. (eds.). *Palaeobiology: a synthesis*. Blackwell Scientific Publications, Oxford, xiii + 583 pp.

Chapter Nine

THE MORRISON FORMATION

Background: terrestrial life in the mid-Mesozoic

From the end of the Triassic to the close of the Cretaceous, life on land was, of course, dominated by the hugely successful dinosaurs. Since different periods of the Mesozoic were characterized by different assemblages of dinosaurs, it is useful to understand something of their classification. Dinosaurs are classified into two main groups, the reptile-hipped saurischians and the bird-hipped ornithischians.

In saurischians the hip bones are arranged similarly to those of most other reptiles. The blade-like upper bone, the ilium, is connected to the backbone by a row of strong ribs and its lower edge forms the upper part of the hip socket. Beneath the ilium is the pubis, which points downward and slightly forward, and behind this is the backwardly extending ischium.

The saurischians are further divided into two groups. Theropods include all of the carnivorous (meat-eating) dinosaurs; most have powerful hind limbs ending in sharply-clawed, bird-like feet, lightly-built arms, a long, muscular tail, and dagger-like teeth. Examples include the giant *Tyrannosaurus* ('tyrant lizard') and *Albertosaurus*, the small and agile *Velociraptor* ('fast-thief'), and some toothless types, such as *Oviraptor* ('egg-thief') and *Struthiomimus* ('ostrich-mimic'), all from the late Cretaceous. The group also includes the classically huge predators of the Jurassic, such as *Allosaurus* and *Megalosaurus* and the tiny *Compsognathus* from the same period (Chapter 10).

The second group of saurischians, the sauropodomorphs, were all herbivorous (plant-eating) dinosaurs. They ranged in size from diminutive forms (the prosauropods, such as *Massospondylus*) from the late Triassic and early Jurassic, to the gigantic true sauropods of the late Jurassic, such as *Diplodocus*, *Apatosaurus* (previously known as *Brontosaurus*), *Brachiosaurus* and *Camarasaurus*. They tended to have long, slender bodies, whip-like tails, long, shallow faces and thin, pencil-shaped teeth.

In the bird-hipped ornithischians the arrangement of the hip bones is similar to that of living birds (although confusingly the ornithischians did not give rise to birds). While the ilium and ischium are arranged in a similar manner to the saurischians, the pubis is a narrow, rod-shaped bone which lies alongside the ischium. In addition, all ornithischians seem to possess a small, horn-covered beak perched on the tip of the lower jaw.

Ornithischians were entirely herbivorous and are classified into five major groups: the ornithopods (medium-sized animals such as the early Cretaceous *Iguanodon* and *Tenontosaurus*, and the hadrosaurs, or duck-billed dinosaurs, such as the late Cretaceous *Edmontosaurus*); the ceratopsians (horned and frilled dinosaurs of the late Cretaceous, such as *Triceratops*); the stegosaurs (plated dinosaurs of the Jurassic, such as *Stegosaurus*); the pachycephalosaurs (with domed and reinforced heads, such as *Pachycephalosaurus*); and the ankylosaurs (armoured dinosaurs, covered in thick bony plates embedded in the skin, such as *Ankylosaurus*).

Fossil evidence for terrestrial life during much of the Jurassic is quite poor, but towards the later part of the period there are some exceptionally rich deposits, especially in China, Tanzania and North America. This chapter is based on the fossils of the Morrison Formation, a vast and highly productive bed, long known for its spectacular dinosaur skeletons, which outcrops along the Front Range of the Rockies from Montana in the north, to Arizona and New Mexico in the south (**153**).

Over such a huge area the Morrison Formation represents a variety of terrestrial conditions from wet swamps (with coal deposits) in the north, to desert conditions in the south. It is in the mid-west states of Colorado, Utah and Wyoming, where the Morrison Formation represents mostly fluviatile and lacustrine deposits, that the richest finds have been made. Here, flash floods deposited literally tons of bones (**154**) in a Concentration Lagerstätte which gives a detailed insight into a late Jurassic terrestrial ecosystem that includes not only some of the largest dinosaurs known, but also some of the other land animals which coexisted alongside the dinosaurs, including the most diverse Mesozoic mammal assemblage yet known.

History of discovery of the Morrison Formation

The story of the discovery of the Morrison Formation dinosaurs has become known in American palaeontological

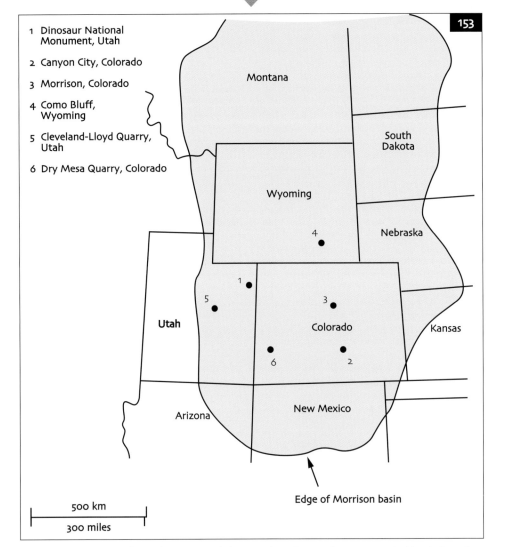

153

1 Dinosaur National
   Monument, Utah

2 Canyon City, Colorado

3 Morrison, Colorado

4 Como Bluff,
   Wyoming

5 Cleveland-Lloyd Quarry,
   Utah

6 Dry Mesa Quarry, Colorado

Montana

South
Dakota

Wyoming

Nebraska

Utah

Colorado

Kansas

Arizona

New Mexico

Edge of Morrison basin

500 km

300 miles

153 Locality map to show the extent of the Morrison Formation outcrop in North America.

folklore as 'The Bone Wars' and is the story of a bitter rivalry between two of America's leading palaeontologists of the late nineteenth century. It began in 1877 when two schoolteachers, Arthur Lakes and O. W. Lucas, independently discovered rich remains of dinosaur bones in Colorado. Lakes sent his fossils, which he had found near the town of Morrison, to Professor Othniel Charles Marsh of Yale's Peabody Museum, who was well known for his work on hadrosaurs from Kansas. In the same year Lucas found bones from the same horizon near Canyon City, which he sent to Edward Drinker Cope at Philadelphia, who had described some of the first ceratopsian dinosaurs from Montana.

Marsh and Cope were already bitter enemies due to an earlier dispute over Cope's description of a fossil reptile which Marsh had shown to be erroneous. Immediately, a frantic race began between the two men to describe the numerous new dinosaurs that were being collected. Cope's specimens from Canyon City were

154

154 Fossil Cabin Museum, Como Bluff, Wyoming. Dinosaur bones are so common that they are used as a building material.

larger and much more complete and at first he had the upper hand, but later the same year (1877) new discoveries were made in equivalent beds at Como Bluff, Wyoming, and this time Marsh was first on the scene.

Como Bluff, near Medicine Bow, Wyoming, is a low ridge, approximately 16 km (10 miles) long and 1.6 km (1 mile) wide formed by a north-east–south-west trending anticline with a gently dipping southern limb and a steeply dipping northern limb. The southern limb is capped by the highly resistant Cloverly Formation of the lowermost Cretaceous, while the northern limb exposes the underlying beds of the uppermost Jurassic, which have since been named the Morrison Formation after the classic locality near Denver. It was in these latter beds that the rich dinosaur fauna, mainly of giant sauropods, was preserved.

The discovery at Como Bluff further fuelled the 'Bone Wars'. It was made by two workers on the trans-continental Union Pacific Railroad, which was then being driven through southern Wyoming to exploit the region's extensive deposits of coal (Breithaupt, 1998). In July 1877 William Edward Carlin, the station agent at nearby Carbon Station, and the section foreman, William Harlow Reed, wrote to Marsh informing him that they had discovered some gigantic bones of what they thought to be *Megatherium* (the Pleistocene ground-sloth), and offered to sell the fossils to Marsh and also to collect further specimens if required. They signed the secret letter with their middle names, Harlow and Edwards, to cover up their identities. Four months later Marsh's representative, Samuel Wendell Williston, arrived at Como Bluff to survey the site and to pay Carlin and Reed their costs.

(A previous cheque made out by Marsh to 'Harlow and Edwards' could not be cashed since these were not their real names!)

Williston immediately informed Marsh of the richness of Como Bluff and transferred his collecting crews from Colorado to the new site. Carlin and Reed continued to work for Marsh, the latter eventually becoming the curator of the Geological Museum of the University of Wyoming in Laramie, and a respected palaeontologist. Not wishing to be outdone, Cope moved his crews to Como Bluff, eventually persuading Carlin to work for him. The feud extended into field operations and skirmishes between rival camps often broke out. Breithaupt (1998) reports spying on rival quarries, smashing of bones to prevent the opposition collecting them, and sometimes even 'fisticuffs'!

Over the ensuing years, however, a huge number of new dinosaurs were discovered including the carnivore *Allosaurus*, the strange, plated dinosaur *Stegosaurus*, and the largest sauropods then known, such as *Diplodocus* and the notorious *Brontosaurus* (now known as *Apatosaurus*). Alongside the dinosaurs the Morrison also yielded the most important fauna of Mesozoic mammals yet discovered.

The rivalry continued until Cope's death in 1897; Marsh died in 1899. By the end of their careers Marsh had described 75 new species of dinosaur, of which 19 are valid today, while Cope had described 55 species of which 9 are valid today. The Morrison Formation, now known from twelve different states in North America (**153**), remains one of the world's most prolific dinosaur 'graveyards' and its fossils can be seen on display in museums all over the world.

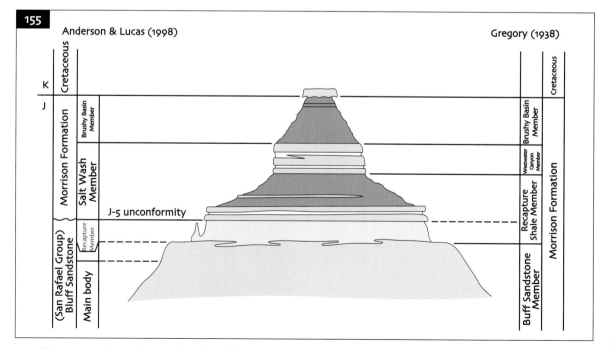

155 Diagram to show the stratigraphy of the Morrison Formation in the region of Bluff, Utah (after Anderson and Lucas, 1998).

Stratigraphic setting and taphonomy of the Morrison Formation

The Morrison Formation, traditionally divided into four members (Gregory 1938), was restricted to a two-member division by Anderson and Lucas (1998): an upper Brushy Basin Member and a lower Salt Wash Member (**155**). It has been dated radiometrically and on microfossil evidence as Kimmeridgian to early Tithonian (Upper Jurassic; approximately 150 million years ago) and its outcrop along the Front Range of the Rocky Mountains (**156**) covers an area of more than 1.5 million square km (0.6 million square miles). The thickness is highly variable, but at Dinosaur National Monument in Utah it is some 188 m (620 ft).

In many areas the Morrison is capped and protected by the highly resistant Cloverly Formation of the Lower Cretaceous (**157**), and is underlain by the Middle Jurassic Sundance Formation representing the marine deposits of the Sundance Sea. This succession, at both Dinosaur National Monument and Cleveland–Lloyd Dinosaur Quarry (Utah), two of the most important Jurassic dinosaur sites in the world, illustrates a regressional sequence coincidental with the final northward withdrawal of this vast, shallow sea during the mid Jurassic. At the latter location, for example, the intertidal beds of the Summerville Formation (equivalent to the Sundance) pass upwards into the supratidal Tidwell Member, then into fluvial and lacustrine deposits of the Salt Wash Member, and finally into overbank deposits with meandering rivers of the Brushy Basin Member (Bilbey, 1998; **155**).

Over such a vast area there is clearly enormous variation in the environment of deposition of the Morrison Formation, but in many of the richest beds the bone accumulations were deposited in poorly sorted sandstones that are thought to have resulted from cataclysmic flash floods. The wide, open plains left by the retreat of the Sundance Sea and traversed by meandering rivers, were home to herds of herbivorous dinosaurs roaming the rivers and lakes in search of vegetation. Cyclic periods of severe drought, with a periodicity of 5, 10, 40 or 50 years as in Kenya today, concentrated the dinosaur herds (and other vertebrates) around remnant water holes where they eventually died of dehydration. Following such mass mortality, subsequent flash floods swept the disarticulated bones a short distance before burial in sandbodies representing the channel fill of streams. This situation was observed at the Dry Mesa Dinosaur Quarry in Colorado by Richmond and Morris (1998), where a diverse assemblage, including 23 different dinosaurs, plus pterosaurs, crocodiles, turtles, mammals, amphibians and lungfish is preserved. The assemblage has gained notoriety for its large sauropods, especially *Supersaurus* and *Ultrasaurus*.

Such mass death assemblages, represented by bone beds in the fossil record, may be either catastrophic or non-catastrophic accumulations. The former is defined as sudden death (within a few hours at most, e.g. a poisonous ash fall) and includes all age and

156 Morrison Formation (foreground) with underlying red beds of the Permo-Triassic resting on Palaeozoic rocks of the Bighorn Mountains (the Front Range of the Rockies); near Buffalo, Wyoming.

157 The Lower Cretaceous Cloverly Formation (top) overlays alternating sandstones and shales of the Morrison Formation bone beds; Wyoming Dinosaur Center, Thermopolis, Wyoming.

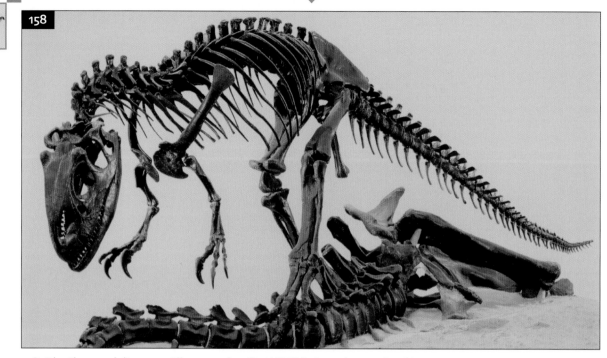

158 The theropod dinosaur *Allosaurus fragilis* (AMNH). Length 12 m (40 ft).

sexual groups. Non-catastrophic mass mortality occurs over a longer time span of hours up to months (e.g. starvation). The killing agent is selective about age, health, gender and social ranking of individuals so that juveniles, females and old adults dominate numerically. Most of the bone accumulations in the Morrison Formation are considered to represent non-catastrophic mass mortalities (Evanoff and Carpenter, 1998) which are characterized by a greater degree of disarticulation due to the longer time span.

The preservation of the disarticulated bones is favoured by an arid climate. Dodson *et al.* (1980) suggested that Morrison Formation dinosaur carcasses decomposed on dry open land or in channel beds prior to deposition. After death the arid climate dehydrated muscle tissue, ligaments and skin, but there is little evidence of scavenging and little sign of cracking or exfoliation of the bones. Richmond and Morris (1998) suggested that they were exposed for no more than 10 years before burial by the flash flood.

Description of the Morrison Formation biota

Allosaurus. This was one of the largest predatory theropod dinosaurs of the Jurassic, up to 12 m (40 ft) in length and weighing up to 1.5 tons (**158–160**). The skull was almost 1 m (3 ft) long, with jaws supporting 70 curved and sharp teeth. It is thought to have hunted in groups when attacking herds of sauropods. The feet supported three ferociously large claws used for tearing flesh.

Diplodocus. Plant-eating sauropod described by Marsh – one of the longest dinosaurs known, up to 27 m

159 The skull of *Allosaurus fragilis*, with sharp serrated teeth for cutting meat (SMA). Length of skull 1 m (3 ft).

(90 ft), but slimly built and weighing only around 10–12 tons (**161–163**). Both the neck and the tail were held more or less horizontally. The 6 m (20 ft) long neck supported a tiny skull compared to the size of the animal, while the long tail with 73 vertebrae could be used like a whiplash. Recent fossil evidence suggests that it possessed a dorsal crest of triangular spikes, like a modern *Iguana*.

160 Reconstruction of *Allosaurus*.

161 The sauropod dinosaur *Diplodocus*; note the whip-like tail (AMNH). Length up to 27 m (90 ft).

162 The sauropod dinosaur *Diplodocus* in matrix (SMA). Length, head to tail, 10.5 m (35 ft).

163 Reconstruction of *Diplodocus*.

Apatosaurus. Another massive plant-eating sauropod related to *Diplodocus* and formerly known as *Brontosaurus.* Around 20 m (65 ft) in length, but with a massive skeleton and weighing more than 20 tons, it was the largest dinosaur known after its discovery by Marsh in the Morrison Formation (**164–166**).

Camarasaurus. Described by Cope, this is the most common herbivorous sauropod and is nicknamed the 'Jurassic cow' (**167–170**). It is related to *Brachiosaurus* (see p. 96), and both had a more upright (giraffe-like) posture than the two previous genera. It was about 18 m (60 ft) long, but had a massive skeleton and weighed up to 18 tons. The skull was much larger than that of *Diplodocus,* supporting 52 cone-shaped teeth for tearing vegetation (**168**).

164 The massive sauropod dinosaur *Apatosaurus* (UWGM). Length 23 m (75 ft).

165 The skull of *Apatosaurus* (see **164**) (UWGM).

166 Reconstruction of *Apatosaurus.*

168 The skull of *Camarasaurus*, with pencil-shaped teeth for tearing vegetation (WDC). Length of skull c. 55 cm (22 in).

167 The sauropod dinosaur *Camarasaurus* (WDC). Length c. 15 m (50 ft).

169 The sacrum of *Camarasaurus* in Morrison Formation bone bed; Wyoming Dinosaur Center, Thermopolis, Wyoming.

170 Reconstruction of *Camarasaurus*.

Stegosaurus. An unusual ornithischian dinosaur with large dorsal plates running in two rows down the length of the body – their function is still not certain, but they may have been used as a temperature regulating device; they consist of only a thin layer of bone and so could not have been used in defence. Also characterized by a tiny skull, front legs shorter than the hind legs, and a massive tail with four sharp spikes – a defensive weapon against large predators (**171–173**).

Dinosaur trace fossils. Ornithopod eggshell, herbivore coprolites and a variety of trackways are all known from the Morrison Formation.

Other reptiles and amphibians. These are never common, but do include some rare records of frogs (the oldest anuran is Lower Jurassic), the lizard-like sphenodons (also known from the Upper Jurassic Solnhofen Limestone; Chapter 10), some true lizards, crocodiles and turtles, and a few records of pterosaurs, including pterodactyloids and rhamphorhynchoids. Records of birds have all been later refuted (Padian, 1998).

Mammals. The Morrison Formation mammals comprise one of the most important Jurassic mammal faunas ever discovered as they provide a rare window on the long early history of mammals in the Mesozoic. Known mainly from isolated jaw bones and teeth are the primitive triconodonts, docodonts, symmetrodonts and dryolestoids, while the multituberculates are a more developed group of rodent-like omnivores which survived into the Eocene (Engelmann and Callison, 1998).

Fish. Lungfish (sarcopterygians) were first reported by Marsh and for many years were the only known Morrison fish. More recently a variety of actinopterygians (ray-finned fish) have been reported including a primitive teleostean (modern bony fish), a variety of holosteans (bony ganoid fish) and a new chondrostean palaeoniscid, *Morrolepis*, the 'Morrison fish' (Kirkland, 1998).

Invertebrates. These include freshwater molluscs (gastropods and bivalves), ostracods, conchostracans, crayfish and caddisfly cases.

Plants. Flora from the Brushy Basin Member includes bryophytes, horsetails, ferns, cycads, ginkgos and conifers (Ash and Tidwell, 1998).

Palaeoecology of the Morrison Formation

The Morrison Formation was deposited in a terrestrial basin near the western margin of Laurasia (following the break-up of Pangaea), situated in the low mid-latitudes between 30° and 40° N. The climate is interpreted as having been arid to semi-arid, but with some seasonal rainfall (Demko and Parrish, 1998). A mountainous region to the west probably had a rain-shadowing effect and low annual rainfall is supported by the presence of evaporites, aeolian sandstones and saline lake facies. However, the presence of various freshwater invertebrates and fish suggests that there were perennial streams and lakes present on the wide, open plains of the Morrison Basin, and the flora of horsetails, ferns, cycads, ginkgos and various gymnosperms suggests at least short periods of a more humid, tropical climate (Ash and Tidwell, 1998). It seems that this fluvial-lacustrine environment was strongly influenced by repeated cycles of drought and flood.

The lush lake margins and swampy river courses were home to huge herds of herbivorous dinosaurs, which roamed the plains in search of food. Smaller quadrupedal herbivores, such as stegosaurs, browsed on low-level horsetails, ferns, cycads and small conifers, while the giant sauropods with their long necks were eating the tops of the tallest trees, mainly conifers, ginkgos and tree ferns. Meat-eating carnivores (such as *Allosaurus*) followed the herbivores and by pack-hunting were able to overcome and kill even the largest sauropod.

Frogs, sphenodons and lizards made their home in and around the lakes and streams which were also inhabited by turtles and crocodiles, the latter being the top predator of these aquatic habitats. Pterosaurs, probably living on lake margins, would also have scanned the lakes in search of fish. Meanwhile another group of small animals was keeping a low profile in caves and in trees, waiting for their day; these small, primitive mammals were mainly rat-like in appearance. Their food would have consisted mostly of insects; although some were true carnivores their prey would of necessity have been small. Most were probably nocturnal (and arboreal) in habit in order to survive the threat of the great carnivorous dinosaurs.

Comparison of the Morrison Formation with other dinosaur sites

Tendaguru Formation, Tanzania

The Upper Jurassic Tendaguru Beds of Tanzania outcrop about 75 km (47 miles) north-west of Lindi, and are the richest deposits of Late Jurassic strata in Africa. German expeditions from 1909 to 1913, led by Werner Janensch and Edwin Hennig, discovered huge accumulations of dinosaur bones comparable in their numbers, age and taxa to the Morrison Formation. Approximately 100 articulated skeletons and many tons of bones were collected and sent to the Museum für Naturkunde in Berlin for study.

The Tendaguru Formation of the Somali Basin differs from the Morrison in having marine horizons. Three members of terrestrial marls are separated by marine sandstones containing Kimmeridgian/Tithonian ammonites. At this time the Tendaguru region was situated between 30° and 40° S. The complex is approximately 140 m (460 ft) thick in total and the depositional regime is interpreted as lagoonal or estuarine within the margin of a warm, epicontinental sea. Russell *et al.* (1980) considered that, as with the Morrison Formation, the dinosaur bones accumulated following mass mortality during periodic regional drought.

The fauna is similar to that of the Morrison in being dominated by giant sauropods, especially the huge *Brachiosaurus*, which was up to 25 m (80 ft) in length and weighed 50–80 tons. The front legs were longer than the hind legs so that it had an upright, giraffe-like posture and stood up to 16 m (50 ft) tall. Although *Brachiosaurus* was first described from fragmentary remains in the Morrison Formation, it is rare in North America and is better known from more complete skeletons from Tendaguru. Other dinosaurs include *Barosaurus* and

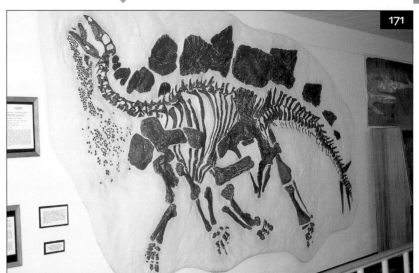

171 The ornithischian dinosaur *Stegosaurus* in matrix (UWGM). Length 4.5 m (15 ft).

172 The ornithischian dinosaur *Stegosaurus,* with dorsal plates and tail spikes (SMA). Length 4.8 m (16 ft).

173 Reconstruction of *Stegosaurus.*

Dicraeosaurus (both diplodocids), *Kentrosaurus* (a stegosaur), *Gigantosaurus* (another huge sauropod), and the small theropod *Elaphrosaurus*.

There are some notable differences between the Morrison and Tendaguru faunas, the most obvious being the rarity of large theropod dinosaurs such as *Allosaurus*. In addition to dinosaurs, the vertebrates also include crocodiles, bony fish, sharks, pterosaurs and mammals. Invertebrates include cephalopods, corals, bivalves, gastropods, brachiopods, arthropods and echinoderms, all inhabitants of the shallow epicontinental sea. A flora of silicified wood plus a microflora of dinoflagellates, spores and pollen may yield new palaeoecological data.

Further Reading

Anderson, O. J. and Lucas, S. G. 1998. Redefinition of Morrison Formation (Upper Jurassic) and related San Rafael Group strata, southwestern U. S. *Modern Geology* **22**, 39–69.

Ash, S. R. and Tidwell, W. D. 1998. Plant megafossils from the Brushy Basin Member of the Morrison Formation near Montezuma Creek Trading Post, southeastern Utah. *Modern Geology* **22**, 321–339.

Bilbey, S. A. 1998. Cleveland-Lloyd dinosaur quarry – age, stratigraphy and depositional environments. *Modern Geology* **22**, 87–120.

Breithaupt, B. H. 1998. Railroads, blizzards, and dinosaurs: a history of collecting in the Morrison Formation of Wyoming during the nineteenth century. *Modern Geology* **23**, 441–463.

Demko, T. M. and Parrish, J. T. 1998. Paleoclimatic setting of the Upper Jurassic Morrison Formation. *Modern Geology* **22**, 283–296.

Dodson, P., Bakker, R. T., Behrensmeyer, A. K. and McIntosh, J. S. 1980. Taphonomy and paleoecology of the dinosaur beds of the Jurassic Morrison Formation. *Paleobiology*, **6**, 208–232.

Engelmann, G. F. and Callison, G. 1998. Mammalian faunas of the Morrison Formation. *Modern Geology* **23**, 343–379.

Evanoff, E. and Carpenter, K. 1998. History, sedimentology, and taphonomy of Felch Quarry 1 and associated sandbodies, Morrison Formation, Garden Park, Colorado. *Modern Geology* **22**, 145–169.

Gregory, H. E. 1938. The San Juan Country. *United States Geological Survey Professional Paper* **188**.

Kirkland, J. I. 1998. Morrison fishes. *Modern Geology* **22**, 503–533.

Padian, K. 1998. Pterosaurians and ?avians from the Morrison Formation (Upper Jurassic, western U.S.). *Modern Geology* **23**, 57–68.

Richmond, D. R. and Morris, T. H. 1998. Stratigraphy and cataclysmic deposition of the Dry Mesa Dinosaur Quarry, Mesa County, Colorado. *Modern Geology* **22**, 121–143.

Russell, D., Béland, P. and McIntosh, J. S. 1980. Paleoecology of the dinosaurs of Tendaguru (Tanzania). *Mémoires de la Société géologique de France* **139**, 169–175.

THE SOLNHOFEN LIMESTONE

Chapter Ten

Background: Mesozoic lithographic limestones (Plattenkalks)

Both marine and terrestrial Jurassic ecosystems have already been analysed in detail (Chapters 8 and 9) from the Lower and Upper Jurassic respectively. However, the geological record of the mid to late Mesozoic includes a disproportionately high number of Fossil-Lagerstätten, due mainly to palaeogeographic conditions during the late Jurassic and early Cretaceous which resulted in a concentration of restricted marine basins at that time.

Many of these basins display a typical facies of very finely laminated, micritic limestones, known as 'lithographic limestones' (some horizons are ideal for lithographic printing) or, more accurately, as 'Plattenkalks', a German word meaning 'platy limestones', which

conveys the idea of the lateral continuity of these beds for many kilometres. For a variety of reasons Plattenkalks often display exquisite preservation of soft tissues. Moreover, Plattenkalks often preserve a more complete ecosystem, including aquatic and terrestrial animals and plants, and thus give a fuller picture than is portrayed, for example, in the more restricted environments of the Holzmaden Shale and the Morrison Formation.

Most celebrated and most important of the various Mesozoic Plattenkalk Lagerstätten is the Solnhofen Limestone of Bavaria in southern Germany (**174**). Although fossils are by no means common, this limestone has produced over the years a range of spectacular specimens which illustrate the richness of life at the very end of the Jurassic. They include the delicate remains of vascular and non-vascular plants, a

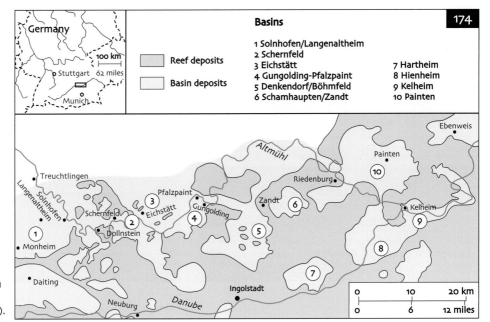

174 Locality map showing the Plattenkalk basins in the Solnhofen Limestone region of the Southern Franconian Alb, Bavaria in southern Germany (after Barthel *et al.*, 1990).

Basins

1 Solnhofen/Langenaltheim
2 Schernfeld
3 Eichstätt
4 Gungolding-Pfalzpaint
5 Denkendorf/Böhmfeld
6 Schamhaupten/Zandt
7 Hartheim
8 Hienheim
9 Kelheim
10 Painten

Reef deposits

Basin deposits

whole range of marine and terrestrial invertebrates (including preservation of the soft tentacles of squids and the fragile wings of dragonflies), fish and marine reptiles, rare dinosaurs, flying reptiles (sometimes complete with wing membranes), and most famous of all, the only known examples of *Archaeopteryx*, the world's earliest known bird, complete with its feathers.

History of discovery and exploitation of the Solnhofen Limestone

There is a long history of exploitation of the Solnhofen Limestone, first as a building stone and later as a lithographic printing stone. The regular bedding and the ease with which it splits along bedding planes into flat blocks or thin sheets have meant that it has been used at least since Roman times for building and for floor and roof tiles. It is still worked today, almost entirely by hand, and produces beautifully coloured floor and wall tiles for domestic use.

At the end of the eighteenth century it was also discovered that certain fine-grained, porous, but hard layers of the Solnhofen Plattenkalk were ideal for lithographic printing, a process that originally required writing in oily ink onto a polished surface of limestone and then etching the exposed limestone in weak acid prior to printing. Hence its common name of 'lithographic limestone'.

The area of outcrop of the Solnhofen Limestone is a high plateau known as the Southern Franconian Alb which lies to the north of Munich in Bavaria (**174**). Outcrop is patchy, with massive biohermal limestones surrounding several distinct Plattenkalk basins. The main quarries are concentrated in the western part of the area, where the limestone is purer, especially around the small village of Solnhofen and the old Baroque town of Eichstätt (**175**, **176**).

Fossils must have been known from these limestones throughout the period of its exploitation, but the real interest began in 1860 when a single fossil feather was discovered near Solnhofen. Further sensation quickly followed when an almost complete skeleton with a fan-like tail and feathered wings, and lacking only the skull, was found the following year (**177**). Although portraying some reptilian features, this was clearly a fossil bird. Its discovery was timely – just two years after Darwin's publication of *The origin of species*, this appeared to be his predicated 'missing link' between reptiles and birds. The specimen, later described as *Archaeopteryx lithographica* (von Meyer, 1861), was given to the local doctor, Carl Häberlein (in lieu of medical fees), who sold it to the British Museum in London (now the Natural History Museum) as part of a collection of Solnhofen fossils.

175 Manual working of the Solnhofen Limestone at Berger Quarry at Harthof, near Eichstätt, in the Southern Franconian Alb of southern Germany.

176 Fine lamination of Plattenkalk beds in the Solnhofen Limestone at Berger Quarry (see **175**).

177 The primitive bird *Archaeopteryx lithographica* – the 'London specimen' (cast in MM; original in NHM). Wingspan 390 mm (15 in).

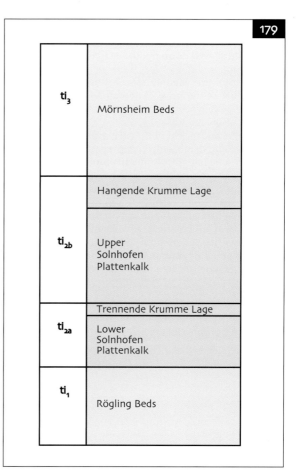

Sixteen years passed before another specimen was found, this time at Blumenberg, near Eichstätt (**178**). The 1877 specimen, complete with skull, was sold to Berlin Museum by Häberlein's son, Ernst. Only five further specimens have ever been discovered: the Maxberg specimen, found in 1955 and since lost; the Haarlem specimen, found in 1855, but only recognized as *Archaeopteryx* in 1970 (in Teylers Museum, Haarlem, Netherlands, where it had been displayed as a pterosaur); the Eichstätt specimen, a juvenile, found in 1950, but only recognized as *Archaeopteryx* in 1973 (in the Jura Museum, Eichstätt, where it had been displayed as the dinosaur *Compsognathus*); the Solnhofen specimen, found in the 1960s (in the Bürgermeister Müller Museum, Solnhofen); and the Solnhofen Aktienverien specimen, found in 1992, and described as a new species, *Archaeopteryx bavarica* (in the Munich Museum).

Stratigraphic setting and taphonomy of the Solnhofen Limestone

The Solnhofen Limestone (strictly the Solnhofen Formation) is divided into a lower and upper Plattenkalk, both dating from the lower part of the Lower Tithonian (Upper Jurassic, zone of *Hybonoticeras hybonotum*, approximately 150 million years ago; **179**). Its outcrop in the Southern Franconian Alb covers an area of 70 km (45 miles) by 30 km (20 miles) and the 95 m (300 ft) sequence is thought to represent a period of about half a million years (Viohl, 1985).

At the beginning of the Jurassic all of the Southern Franconian Alb lay under a wide shelf sea which by middle Jurassic times had connections with the Tethys Ocean to the south (Barthel *et al.*, 1990, fig. 2.6). It is generally accepted that the Solnhofen Limestone was deposited in a restricted lagoon within this shelf sea, protected by the Mitteldeutsche Schwelle landmass to the north (see Chapter 4, p. 40) and by coral reefs to the south and east, separating the lagoon from the Tethys Ocean.

By the late Jurassic (Oxfordian Stage), sponges and cyanobacteria (blue-green 'algae') began to build reef mounds in this lagoon and micritic lime mud was deposited in basins between them. From the Kimmeridgian onwards the basinal deposits included Plattenkalks, and in the early Tithonian the Solnhofen Plattenkalk was deposited. The regular lamination of the Plattenkalks testifies to their deposition in quiet, protected waters (compare with Burgess Shale, Chapter 2, and Holzmaden Shale, Chapter 8).

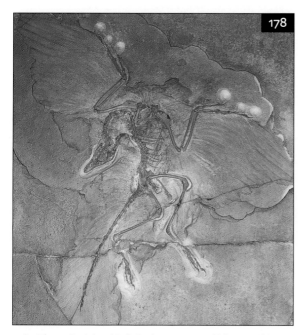

178 The primitive bird *Archaeopteryx lithographica* – the 'Berlin specimen' (cast in MM; original in HMB). Wingspan 430 mm (17 in).

179 Diagram to show the stratigraphy of the Upper Jurassic succession in the Solnhofen–Eichstätt area (after Barthel *et al.*, 1990).

Such restricted lagoonal waters would have been stagnant, and high evaporation rates under the semi-arid climate led to salinity density stratification with heavy, hypersaline bottom waters hostile to life (Viohl, 1985, 1996). The toxic bottom waters may also have been anoxic (lacking oxygen) and/or further poisoned by algal blooms. The lateral continuity over large distances of the thin Plattenkalk beds, with no evidence of bioturbation of the sediment by crawling or burrowing organisms, suggests a lack of benthic fauna in the lagoon. Surface waters, on the other hand, were probably better aerated with normal salinities, allowing some planktonic (floating) and nektonic (swimming) organisms (as evidenced by their coprolites), while both nektonic and benthic forms may have survived on top of the sponge–microbial mounds which reached the oxygenated surface waters (Viohl, 1996).

The depositional model proposed by Barthel (1964, 1970, 1978) suggests that violent monsoonal storms periodically caused a mixing of surface and bottom waters with sudden fluctuations in salinity causing the death of any organisms living near the surface. Fossil evidence does point to mass mortality events and of sudden death (for example, fish killed while in the process of eating). Marine faunas were also washed into the lagoon from the reef community and from the open sea, airborne pterosaurs and *Archaeopteryx* were caught in high winds and drowned, while flying insects and plant fragments were blown across the lagoon and sank. The stagnant, hypersaline conditions excluded scavengers and slowed down microbial decay of the corpses. Some animals did live for short periods after being washed into the lagoon (for example, the famous 'death trails' of the horseshoe crab *Mesolimulus* and the crustacean *Mecochirus*, whose bodies are preserved at the ends of their trails).

The storms played a further part by stirring up carbonate ooze that had been deposited around the coral reefs and washing this sediment into the lagoon. The finer particles, re-suspended in the turbulent water, were carried north to the Plattenkalk basins where they were finally deposited out of suspension, rapidly burying any corpses that had fallen to the lagoon floor. (In this model the sediment is regarded as allochthonous. Keupp's depositional model [Keupp, 1977a,b] differs slightly in regarding the sediment as autochthonous, produced by cyanobacteria on the lagoon floor.)

Rapid burial ensured that intricate details of soft tissue, such as the wings of insects, the tentacles of squids and the feathers of birds, were preserved in the fine mud as simple impressions. Occasionally organic material is preserved unaltered, such as the ink-sacs of cephalopods or as in the original single feather of *Archaeopteryx*, or may be replaced by calcium phosphate (francolite), most usually in the muscles of fish and cephalopods. A cyanobacterial mat on the lagoon floor (suggested by the presence of hollow spheres of coccoid cyanobacteria in the sediment) may have played a further role in this preservation by encapsulating corpses and by binding together the carbonate ooze, preserving tracks and traces.

Description of the Solnhofen Limestone biota

Archaeopteryx. The earliest known bird in the fossil record displays a number of reptilian characters (**177, 178, 180**). The hands have three sharply clawed fingers that are not incorporated into the wings, the jaws are edged with sharp teeth, and it has a long, bony, reptilian tail. Avian features include long, slender legs with bird-like feet, a tail fringed by a fan of feathers, a strong furcula (wishbone) near the front of the chest, and wings with asymmetric flight feathers. The latter feature suggests that *Archaeopteryx* was able to fly (Feduccia and Tordoff, 1979). However, Ostrom (1974) suggested that it would not have been an efficient flier as the sternum lacked a keel for the attachment of the pectoralis muscles and the coracoids lacked attachment processes for the supracoracoideus muscles, which lift the wing. Ostrom (1985) thus regarded it as 'a feeble flapper'. As to its lifestyle Martin (1985) and Yalden (1985) argued for an arboreal habitat, the latter showing that the claws of the manus had a climbing function as with treecreepers and woodpeckers.

Compsognathus. The only dinosaur from the Solnhofen Limestone was a tiny, chicken-sized coelurosaurian theropod (**181, 182**). The long neck and small, swivelling head were balanced by a long tail. Long, powerful hindlimbs contrast with short forelimbs with two-fingered hands. The skeleton of this bird-like dinosaur has many features in common with *Archaeopteryx*. The single Solnhofen specimen contains the skeleton of a lizard in its stomach. A second specimen was discovered in 1972 from the Tithonian of Provence in France.

180 Reconstruction of *Archaeopteryx*.

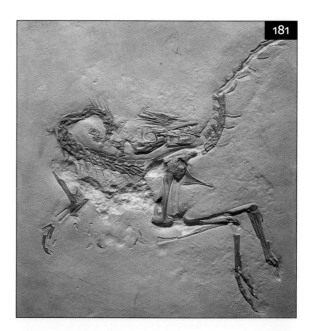

181 The small theropod dinosaur *Compsognathus longipes* (cast in MM; original in BSPGM). Width of block 310 mm (c. 12 in).

182 Reconstruction of *Compsognathus*.

Pterosaurs. Solnhofen pterosaurs were generally small compared to the Cretaceous giants of the Santana Formation (Chapter 11). They include 'long-tailed' rhamphorhynchoids, such as *Rhamphorhynchus* (**183**) and *Scaphognathus*, with wingspans up to 1 m (3 ft), and 'short-tailed' pterodactyloids, such as the thrush-sized *Pterodactylus* (**184**). Wing membranes and the rudder-like tail membranes may be preserved as impressions sometimes showing hair-like coverings, and some specimens show webbing between the toes.

Other reptiles. These include ichthyosaurs, plesiosaurs, crocodiles, turtles, lizards and sphenodons. Ichthyosaurs and plesiosaurs (known from a single tooth) are rare and always poorly preserved. Such strong swimmers from the open sea would only be washed into the lagoon in severe storms. Lizards and lizard-like sphenodons are also uncommon, suggesting that they lived inland far from the lagoon shores. Turtles include freshwater/coastal types and crocodiles include marine and terrestrial forms, but again they are rare in the Plattenkalks.

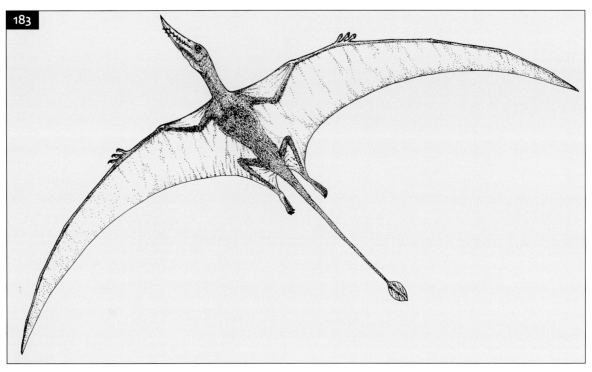

183 Reconstruction of *Rhamphorhynchus.*

184 Reconstruction of *Pterodactylus.*

Fish. The most common Solnhofen vertebrates include actinopterygians (ray-finned fish), sarcopterygians (lobe-finned fish) and chondrichthyans (cartilaginous fish). Actinopterygians are mostly holosteans (the bony ganoid fish) such as *Lepidotes* (**146**), up to 2 m (6 ft) long, *Gyrodus*, resembling a parrot fish, and *Caturus*, a giant predator, but some of the early teleosteans (modern bony fish) are also known, such as the sprat-like *Leptolepides* (**185**), commonly found in mass mortality assemblages. Sarcopterygians are not common, but include the small coelacanth *Coccoderma*, while the chondrichthyans include sharks, rays and chimaeras (ratfish).

Crustaceans. Decapod crustaceans (shrimps, lobsters and crabs) are perhaps the best known of Solnhofen's marine invertebrates and common genera include *Aeger* (**186**, **187**), *Mecochirus*, and *Cycleryon*. *Mecochirus* and the horseshoe crab *Mesolimulus* are commonly found fossilized at the end of a spiral or haphazard trail, suggesting that they landed on the toxic lagoon bottom, struggled for a few steps and then died (**188**).

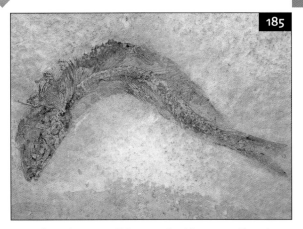

185 The teleostean fish *Leptolepides sprattiformis* (MM). Length 75 mm (3 in).

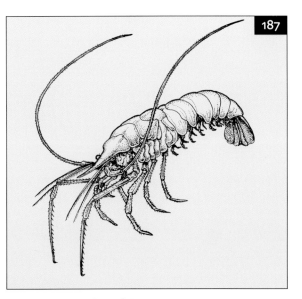

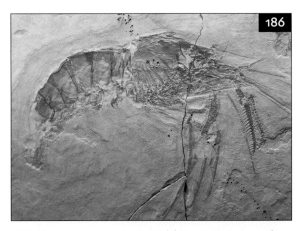

186 The crustacean *Aeger tipularius* (MM). Length 65 mm (2.5 in).

187 Reconstruction of *Aeger*.

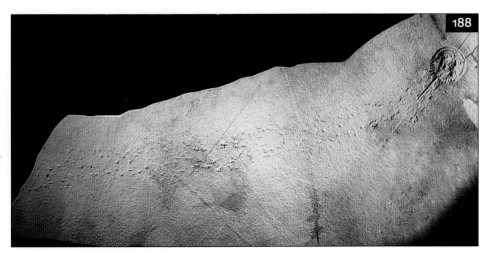

188 The horseshoe crab *Mesolimulus*, preserved at the end of its death trail (PC). Length of trackway 1 m (3 ft).

Insects. Terrestrial arthropods are usually preserved as impressions; only winged insects are known, but fine detail of wing venation is often preserved. They include mayflies, dragonflies (**189**), cockroaches and termites, water skaters, locusts and crickets, bugs and water scorpions, cicadas, lacewings, beetles, caddis flies, true flies and wasps.

Other invertebrates. Most marine invertebrate groups are represented including sponges, jellyfish, corals, annelids, bryozoans, brachiopods, bivalves, gastropods, cephalopods (squids, belemnoids, nautiloids and ammonoids) and echinoderms (crinoids, starfish, brittle stars, sea urchins and sea cucumbers). The squids (such as *Acanthoteuthis*) often have their tentacles and tentacular hooks preserved as impressions (**190**, **191**), and their ink sacs preserved as original carbon. The floating crinoid *Saccocoma* is one of the Solnhofen's most common fossils (**192**).

Plants. All the vascular plants from Solnhofen are gymnosperms. They include pteridosperms (seed ferns), bennettitales, ginkgos and conifers, but there is no evidence of large trees.

Trace fossils. Trace fossils are important in the Solnhofen Limestone and include coprolites, such as the worm-like *Lumbricaria*, which includes disaggregated plates of floating crinoids. Different ichnospecies of *Lumbricaria* probably represent the faeces of fish and squids and indicate that the lagoon was not entirely devoid of autochthonous life. Settling traces, of ammonites for example, show that some bottom waters were stagnant, while drag marks suggest weak currents in some places. The trails of crustaceans have already been described.

The fauna and flora from Solnhofen is well documented and illustrated by Frickhinger (1994).

189 The dragonfly *Tarsophlebia eximia* (MM). Wingspan 75 mm (3 in).

190 The squid *Acanthoteuthis* sp. (MM). Length 300 mm (12 in).

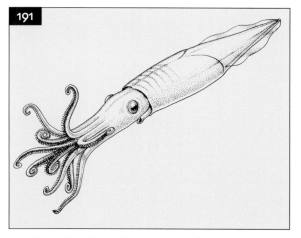

191 Reconstruction of *Acanthoteuthis*.

192 The floating crinoid *Saccocoma tenellum* (MM). Width 35 mm (1.4 in).

Palaeoecology of the Solnhofen Limestone

The Solnhofen Limestone Plattenkalks represent a shallow, saline lagoon community, situated within the subtropical zone at about 25°–30°N, with a semi-arid monsoonal climate. Stagnant, hypersaline bottom waters inhibited colonization or scavenging and there was no autochthonous benthic life except for the salt-loving cyanobacterial mat.

Better-aerated surface waters with normal salinities may have supported limited planktonic and nektonic life, at least for short periods following episodes of water mixing. For example, floating crinoids are found in large numbers at some horizons, and these were clearly prey to both fish and squids within the lagoon whose coprolites containing crinoid remains were deposited on the lagoon floor. Viohl (1996) identifies four habitats occupied by fish, and includes the oxic surface waters of the lagoon as well as the tops of the sponge–microbial mounds. The latter projected upwards into the equitable surface waters and some benthic fauna, especially crustaceans, also lived on these tops.

The inhospitable lagoon was separated from the open sea by a series of coral reefs which supported a rich marine fauna of invertebrates, fish and marine reptiles. The Solnhofen biota is biased towards those forms most likely to be washed into the lagoon by storms (Barthel *et al.*, 1990, p. 89). From the reef community planktonic floaters, such as crinoids, jellyfish and oysters attached to seaweed, were accompanied by weak swimmers such as ammonites and small fish which swam around the reef. Less common are the stronger swimmers, such as squids, large predatory fish, and marine reptiles (ichthyosaurs, plesiosaurs and crocodiles) all of which inhabited the open sea. Vagrant benthos such as crustaceans, horse-shoe crabs, gastropods, echinoids and starfish, were more likely to be washed in than the sessile benthos. Thus, the benthic epifauna, such as sponges, corals, bryozoans, brachiopods, and attached bivalves (e.g. *Pinna*), and the benthic infauna, including burrowing bivalves (e.g. *Solemya*) and polychaete worms (e.g. *Ctenoscolex*), are often represented only by fragments.

An adjacent low-lying landmass to the north of the lagoon supported an abundance of shrubby vegetation of low conifers and other gymnosperms adapted to salty soil. There were no large trees. Amongst this vegetation lived a variety of aquatic, semi-aquatic and terrestrial insects, which would have been a rich source of food for small lizards, the beaked rhynchocephalians and birds (*Archaeopteryx*). Bird-like theropod dinosaurs in turn fed on the lizards.

Coastal waters were home to turtles and crocodiles, while the shores were inhabited by numerous ptero-saurs, mostly in search of fish (as evidenced by stomach contents) although some were possibly insectivores (suggested by their teeth). Again the Solnhofen biota is biased, the majority of the terrestrial component being flying forms, either insects, pterosaurs or birds, which could traverse the lagoon. Land-based animals, such as the lizards and dinosaurs, are understandably extremely rare.

More than 600 species have been described from the Solnhofen Limestone, representing a number of different environments. Perhaps the most striking feature is that the vast majority of these are allochthonous, having been swept into the lagoon from the reef community, the open sea or from the adjacent land.

Comparison of the Solnhofen Limestone with other mid-Mesozoic biotas
Liaoning Province, north-east China

There are several mid-Mesozoic Lagerstätten com-parable to Solnhofen (see Chapter 11), but in terms of ecosystems it is interesting to consider the recently discovered deposit in Liaoning Province in China. Large quarries have been excavated by local farmers in search of fossils, near the village of Sihetun to the south of Beipiao City in western Liaoning, and vast collections have been made of a varied biota with soft-tissue preservation.

The Chaomidianzi Formation (previously referred to as the lower section of the Yixian Formation; see Chiappe *et al.*, 1999) is of late Jurassic or early Cretaceous age, and has been made famous since the discovery in 1994 (Hou *et al.*, 1995) of the first toothless, beaked bird, *Confuciusornis sanctus*, and more so since the discovery two years later of 'feathered dinosaurs' from the same beds (Ji *et al.*, 1999).

The varied bird fauna, with at least eight described genera and many more awaiting description, is un-doubtedly the oldest in the fossil record after the Solnhofen *Archaeopteryx*. The birds all display preser-vation of feathers and many of their morphological features suggest that they had reached a more advanced evolutionary stage than *Archaeopteryx*; for example, *Confuciusornis* had a toothless beak, had lost the bony reptilian tail, and possessed a backward-pointing first toe (hallux) – an early adaptation for perching.

The biota is varied and includes freshwater molluscs (gastropods and bivalves); ostracods, conchostracans and shrimps; fish, frogs, turtles, lizards, pterosaurs and small mammals; psittacosaurid dinosaurs, dromaeo-saurs (e.g. *Sinornithosaurus*), a therizinosaur (*Beipiao-saurus*) and the feathered dinosaurs (*Sinosauropteryx*, *Caudipteryx* and *Protarchaeopteryx*); plus several birds including *Confuciusornis*. There is also a diverse insect fauna (dragonflies, stick insects, butterflies and moths, snakeflies, lacewings, true flies, planthoppers and grasshoppers), occasional spiders, and a flora including cycads, ginkgos, conifers and angiosperms.

The Chaomidianzi Formation is divided into three members, all of which consist of thinly bedded tuffs and/or tuffaceous sediments, but the precise age of the succession is unresolved. Radiometric dates from intrusive igneous rocks suggest an Early Cretaceous age, as do palynological analysis and the presence of the ornithischian dinosaur *Psittacosaurus*. However, radio-metric dates from the tuffaceous sediments themselves and the occurrence of a rhamphorhynchoid ('long-tailed') pterosaur (Ji *et al.*, 1999) suggest assignment to the latest Jurassic.

The succession represents a transition from fluvial to lacustrine, followed by the gradual filling of the lake. Intense volcanic activity, with the production of large volumes of ash, possibly led to a succession of mass mortality events which affected the entire lake ecosystem (Viohl, 1997). Carcasses settling on the lake bed, both from within the lake and from adjacent land areas, remained undisturbed as potential scavengers had also been killed, and would have been rapidly buried by the fine ash. Corroboration of these hypotheses awaits detailed taphonomic research, but episodic production of large volumes of fine sediment recalls the mud flows of the Burgess Shale (Chapter 2) and the storm deposits of Solnhofen (this Chapter), and is responsible for the fine lamination of the sediments.

Comparison of Liaoning to Solnhofen is interesting; the fact that this is a freshwater lake rather than a lagoon dictates that the rich marine fauna of Solnhofen is replaced by freshwater fish and invertebrates, but much of the allochthonous biota is remarkably similar. Both include birds, pterosaurs, small dinosaurs, turtles and lizards; both include a diverse fauna of insects and a flora of ginkgos and conifers. Angiosperms are not present at Solnhofen.

Further Reading

Barthel, K. W. 1964. Zur Entstehung der Solnhofener Plattenkalke (unteres Untertithon). *Mitteilungen der Bayerische Staatssammlung für Paläontologie und Historische Geologie* **4**, 37–69.

Barthel, K. W. 1970. On the deposition of the Solnhofen lithographic limestone (Lower Tithonian, Bavaria, Germany). *Neues Jahrbuch für Geologie und Paläontologie, Abhandlungen* **135**, 1–18.

Barthel, K. W. 1978. *Solnhofen: Ein Blick in die Erdgeschichte*. Ott Verlag, Thun, 393 pp.

Barthel, K. W., Swinburne, N. H. M. and Conway Morris, S. 1990. *Solnhofen; a study in Mesozoic palaeontology*. Cambridge University Press, Cambridge, x + 236 pp.

Chiappe, L. M., Ji, S., Ji Q. and Norell, M. A. 1999. Anatomy and systematics of the Confuciusornithidae (Theropoda: Aves) from the late Mesozoic of northeastern China. *Bulletin of the American Museum of Natural History* **242**, 1–89.

Feduccia, A. and Tordoff, H. B. 1979. Feathers of *Archaeopteryx*: asymmetric vanes indicate aerodynamic function. *Science* **203**, 1021–1022.

Frickhinger, K. A. 1994. *The fossils of Solnhofen*. Goldschneck-Verlag, Weidert, 336 pp.

Hou, L.-H., Zhou, Z.-H., Gu, Y.-C. and Zhang, H. 1995. *Confuciusornis sanctus*, a new Late Jurassic sauriurine bird from China. *Chinese Scientific Bulletin* **40**, 1545–1551.

Ji, Q., Currie, P., Ji, S.-A. and Norell, M. A. 1999. Two feathered dinosaurs from northeastern China. *Nature* **393**, 753–761.

Ji, S.-A., Ji, Q. and Padian, K. 1999. Biostratigraphy of new pterosaurs from China. *Nature* **398**, 573–574.

Keupp, H. 1977a. Ultrafazies und Genese der Solnhofener Plattenkalke (Oberer Malm, Südliche Frankenalb). *Abhandlung der Naturhistorischen Gesellschaft Nürnberg* **37**.

Keupp, H. 1977b. Der Solnhofener Plattenkalk – ein Blaugrünalgen-Laminit. *Paläontologische Zeitschrift* **51**, 102–116.

Martin, L. D. 1985. The relationship of *Archaeopteryx* to other birds. 177–183. *In* Hecht, M. K., Ostrom, J. H., Viohl, G. and Wellnhofer, P. (eds.). *The beginnings of birds*. Proceedings of the International *Archaeopteryx* Conference, Eichstätt, 1984. 382 pp.

Meyer, H. von. 1861. *Archaeopteryx lithographica* (Vogel-Feder) und *Pterodactylus* von Solnhofen. *Neues Jahrbuch für Mineralogie, Geologie und Paläontologie* **1861**, 678–679.

Ostrom, J. H. 1974. *Archaeopteryx* and the origin of flight. *Quarterly Review of Biology* **49**, 27–47.

Ostrom, J. H. 1985. The meaning of *Archaeopteryx*. 161–176. *In* Hecht, M. K., Ostrom, J. H., Viohl, G. and Wellnhofer, P. (eds.). *The beginnings of birds*. Proceedings of the International *Archaeopteryx* Conference, Eichstätt, 1984. 382 pp.

Viohl, G. 1985. Geology of the Solnhofen lithographic limestone and the habitat of *Archaeopteryx*. 31–44. *In* Hecht, M. K., Ostrom, J. H., Viohl, G. and Wellnhofer, P. (eds.). *The beginnings of birds*. Proceedings of the International *Archaeopteryx* Conference, Eichstätt, 1984. 382 pp.

Viohl, G. 1996. The paleoenvironment of the Late Jurassic fishes from the southern Franconian Alb (Bavaria, Germany). 513–528. *In* Arratia, G. and Viohl, G. (eds.). *Mesozoic fishes – systematics and paleoecology*. Proceedings of the international meeting, Eichstätt, 1993. Dr Friedrich Pfeil Verlag, Munich, 576pp.

Viohl, G. 1997. Chinesische Vögel im Jura-Museum. *Archaeopteryx* **15**, 97–102.

Yalden, D. W. 1985. Forelimb function in *Archaeopteryx*. 91–97. *In* Hecht, M. K., Ostrom, J. H., Viohl, G. and Wellnhofer, P. (eds.). *The beginnings of birds*. Proceedings of the International *Archaeopteryx* Conference, Eichstätt, 1984. 382 pp.

THE SANTANA AND CRATO FORMATIONS

Chapter Eleven

Background: the break-up of the Pangaea supercontinent

The onset of the Cretaceous Period saw the super-continent of Pangaea continue to break up, owing to movements deep in the Earth's mantle. Initially North America and Europe pulled apart, opening up the North Atlantic Ocean, and by the end of the early Cretaceous South America and Africa also began to cleave apart to form the South Atlantic. Meanwhile the Tethys Ocean extended westwards so that the northern continents were divided from those in the south. This loss of land bridges and migration routes led to a diversification of animal and plant life on separated continents.

During this time the Earth was warmer than at any other time in its history – atmospheric CO_2 was high and the greenhouse effect led to warm, dry climates. Polar ice caps melted and the rise in sea level flooded large areas of the continents with shallow seas. At the end of the early Cretaceous intense volcanic activity, particularly at mid-ocean ridges, lifted ocean floors and sea levels rose even further. The volcanic activity released more CO_2, further exaggerating the green-house effect.

The dominance of large reptiles on land (dinosaurs), in the sea (ichthyosaurs and plesiosaurs) and in the air (pterosaurs), evident during the Jurassic (Chapters 8, 9, 10), continued into the Cretaceous, but this period witnessed significant evolutionary innovations in all these groups before their final extinction at the Cretaceous/Tertiary (K/T) boundary.

As the continents divided, dinosaur groups began to diverge. While at the start of the Cretaceous many dino-saurs (such as *Iguanodon*) were able to migrate between North America and Europe, by the middle of the period, although *Iguanodon* continued to thrive in Europe, the dominant ornithopod (bipedal plant eater) in North America was *Tenontosaurus*. In South America the massive sauropods (giant plant eaters), so successful in the Jurassic (see Chapter 9), continued to dominate, but in northern continents these and the stegosaurs (plated dinosaurs) quickly declined, their place taken by smaller, ornithischian herbivores, which moved in vast herds across the Cretaceous landscape. They included the ankylosaurs (armoured dinosaurs) and the first ceratopsians (horned dinosaurs, such as *Psittacosaurus*, *Protoceratops* and *Triceratops*). At the same time the iguanodonts gave rise to the duck-billed hadrosaurs, such as *Maiasaura*.

Amongst the carnivores the allosaurs (see Chapter 9) began to decrease and were replaced by huge thero-pods, such as *Tyrannosaurus*, and smaller predators such as *Deinonychus*, *Velociraptor* and the ostrich-like *Ornithomimus*.

In the water the dominant predators were massive marine lizards, the mosasaurs, which had replaced the ichthyosaurs of the Jurassic. Plesiosaurs continued to thrive, along with giant turtles, but marine crocodiles disappeared early in the Cretaceous. Fish evolved rapidly, but the more primitive holostean bony fish of the Jurassic were gradually replaced by the teleostean bony fish, and by the end of the Cretaceous most holosteans had disappeared. Early teleosts were primitive, including herring-like and salmon-like ancestors. The ancient hybodont sharks made their final appearance in the Cretaceous. Their advanced shark descendants were numerous and included many modern forms; skates and rays had also essentially reached modern conditions. *Mawsonia* is the last known fossil coelacanth and was thought to be the last representative of the group until the discovery in 1938 of a living representative.

In the air the primitive 'long-tailed' pterosaurs (the rhamphorhynchoids), dominant in the Jurassic, were replaced by the 'short-tailed' pterodactyloids, some with wingspans up to 15 m (50 ft) – the pinnacle of pterosaur evolution. But from the beginning of the Cretaceous they had to share the airways with their first com-petitors, the birds, which developed from feathered dinosaurs during the late Jurassic and earliest Cretaceous. Initially, birds seemed to have thrived around lakes, but due to their greater adaptability than pterosaurs they soon spread to a variety of habitats.

However, arguably the most important evolutionary event of the Mesozoic was the appearance of angiosperms, the flowering plants. The angiosperms had developed a new strategy to defend themselves against grazing animals, by growing and reproducing faster. As the Cretaceous Period progressed, flowers developed a closer relationship with insects. Flowers needed to ensure that the pollen grain reached the protected ovule and so they began to produce nectar to attract insects, which carried excess pollen from one plant to another. As the angiosperms underwent explosive growth at the end of the early Cretaceous, so the insects diversified rapidly in parallel development and the Cretaceous air was full of flying reptiles, birds and insects.

Much of our knowledge of the fauna and flora during this Cretaceous Period of innovative evolution comes from one of the world's most productive fossil sites, located in the State of Ceará in north-eastern

Brazil (**193**). There, in a succession 700 m (2,300 ft) thick of mainly Lower Cretaceous sediments, can be found not one, but two separate Fossil-Lagerstätten which, by modern stratigraphical nomenclature, are now separately named as the Santana Formation and Crato Formation.

Soft-tissue preservation by very different mechanisms has provided an insight into a variety of early Cretaceous biotas. In the Santana Formation these include a variety of fish, spectacular pterosaurs and other reptiles, including dinosaurs, all preserved inside limestone concretions, while the Crato Formation displays the world's most remarkable Cretaceous insect fauna, a diverse flora of gymnosperms and early angiosperms, a number of fish, and rare reptiles and amphibians, all preserved in a micritic Plattenkalk limestone, not dissimilar to that of Solnhofen (Chapter 10).

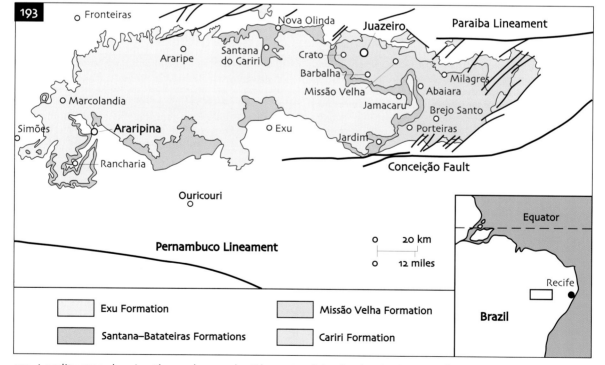

193 Locality map showing the geology and settlements of the Araripe Basin in north-eastern Brazil (after Martill, 1993).

History of discovery of the Santana and Crato Formations

The discovery of the remarkable fossils of the Santana Formation goes back to Napoleon's destruction of the Portuguese Empire in the early nineteenth century, which led to Portugal's renewed interest in the exploitation of Brazil. Don Pedro, Brazil's first Portuguese emperor, married an Austrian archduchess, which led to an influx of scientists and philosophers coming to Brazil from Austria and Bavaria.

In 1817 and 1820 the German naturalists Johann Baptist von Spix and Carl Friedrich Philipp von Martius, from the Academy of Sciences in Munich, explored much of Brazil, including the north-eastern state of Ceará, where they came across the fish-bearing concretions from what was later described as the Santana Formation. Their report was published between 1823 and 1831 and includes the first illustration of a Santana fish.

The fossils are found in sediments exposed around the flanks of an uplifted, flat-topped plateau, known as the Chapada do Araripe, at about 800 m (2,600 ft) elevation. The Chapada represents the remnants of an ancient sedimentary basin, the Araripe Basin, and trends east–west, straddling the states of Ceará to the north, and Pernambuco to the south (**193**).

This region was visited again during the years 1836–41 by Glasgow botanist George Gardner, whose book, *Travels in the interior of Brazil* (1846), gives a fascinating account of his collection of fossil fishes from 'rounded limestones'. These were sent back to the UK to be exhibited at the British Association in Glasgow, where they were seen by the eminent palaeontologist Louis Agassiz, who described seven species and correlated them to the Cretaceous Period (Agassiz, 1841; Gardner, 1841).

Jordan and Branner's 1908 monograph was the first major palaeontological work on the Santana fish, while Small (1913) gave the first stratigraphical account of the Araripe Basin, and introduced the name Sant' Ana Limestones. Beurlen's more recent stratigraphic review (1962) divided the Santana Formation into three members, a lower Crato Member, consisting of shales and laminated limestones (with insects and plants), and an upper Romualdo Member, containing the characteristic fish concretions, these being separated by the Ipubi Member of evaporites.

There has been much controversy over the stratigraphic nomenclature of this sequence, but that suggested by Martill (1993) – in which the Santana Formation is now restricted to the concretion-bearing sequence previously described as the Romualdo Member, while the Ipubi and Crato members are both elevated to formational status (**194**) – seems to be acceptable.

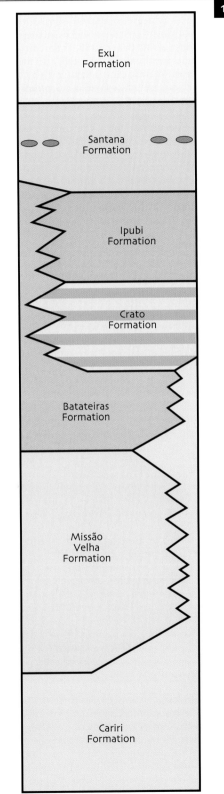

194 Diagram to show the stratigraphy of the Mesozoic succession in the Araripe Basin (after Martill, 1993).

Besides the well-known fish, the Santana Formation (*s.s.*) has recently begun to yield fossil reptiles including crocodiles, turtles, dinosaurs, and spectacular pterosaurs, all found within the concretions. The concretions are 'mined' by peasant farmers (**195–198**), especially around the villages of Cancau (near Santana do Cariri) and Jardim, and are sold cheaply to local commercial fossil dealers, even though trade in fossils is illegal in Brazil.

The spectacular insects and plants of the Crato Formation limestones have only really been known since the 1980s (see Grimaldi, 1990). They have been mostly discovered by the workers, many of them children, who quarry by hand the laminated limestones around the village of Nova Olinda for use as an ornamental paving stone (**199, 200**). Fossils from both the Santana and Crato formations were beautifully illustrated by Maisey in his 1991 book, *Santana fossils.*

Stratigraphic setting and taphonomy of the Santana and Crato Formations

The Araripe Basin, in which these formations occur, is a fault-bounded interior basin whose evolution is closely related to rifting associated with the opening in Lower Cretaceous times of the South Atlantic Ocean, brought about by South America and Africa pulling apart. The Proterozoic crystalline basement of Brazil is composed of a fractured craton and many of the faults were probably reactivated during the break-up of Pangaea to form coastal and interior sedimentary basins. The Araripe Basin lies between two major east-west trending lineaments (the Paraiba and Pernambuco lineaments; see Martill, 1993, text-figs. 2.2, 2.3, 2.4), which can be traced across the mid-Atlantic Ridge as transform faults and correlated with their corresponding continental fault zones in west central Africa.

An erosional remnant of the basin has since been uplifted to form the 200 km (125 miles) wide, east–west trending plateau of the Chapada do Araripe, in which horizontal Cretaceous (and possibly also Jurassic) sediments lie unconformably on the Proterozoic/Palaeozoic basement (**193**). As noted in the previous section there has been much discussion on the stratigraphy of the Araripe Basin, but the scheme suggested by Martill (1993), which divides the Mesozoic succession into seven formations, seems sensible as it demonstrates the complex interrelationships between the formations which previous workers have ignored (**194**).

The four youngest formations are placed by Martill (1993) in the Araripe Group, which includes all the important fossil-bearing beds. At the base of this Group the Crato Formation consists mainly of laminated, micritic limestones (Plattenkalks) approximately 30 m (100 ft) thick. It comprises three members, and it is the basal Nova Olinda Member which contains the well-preserved insects and plants. The overlying Ipubi Formation, consisting of bedded evaporites, mainly gypsum, up to 20 m (65 ft) thick, is devoid of fossils.

Above the evaporites, the Santana Formation consists of non-fluvial, deltaic silts and sands which give way to a series of green/grey laminated shales with fossil-bearing concretions. The upper part of the formation, above the level of the concretions, consists of shales with thin limestones containing gastropods and rare echinoids. The overlying Exu Formation, 75 m (250 ft) of coarse, cross-bedded sandstones, forms a resistant, horizontal cap to the Chapada plateau, with the older formations exposed on the plateau flanks.

The exact age of the Crato and Santana Lagerstätten within the Lower Cretaceous has yet to be determined, partly due to the fact that many of the fossils have been collected by local people who have not recorded their precise provenance. Based on preliminary palynological results the Crato Formation is thought to be of Upper Aptian/Lower Albian age, while the Santana Formation is probably of Albian age.

Although of similar age, the mechanisms of soft part preservation of these two biotas were quite distinct. In the Crato Formation the preservation of the insects is exceptional, with microstructural detail and even colour patterns preserved (Martill and Frey, 1995). Scanning electron microscopy has revealed eye facets and fine hairs on cuticle. Most of the plants and insects have been pyritized and oxidized to goethite, with no original carbon preserved.

The environment of deposition of the Crato Formation Plattenkalks is generally considered to have been a freshwater lake developed within the interior basin, but in which salinity was continually increasing due to the arid climate. A salinity-stratified water column and/or oxygen-deficient stagnant bottom waters would have prevented any autochthonous life except in the freshwater tongues developed at lake margins around river mouths. The small freshwater fish *Dastilbe* probably lived in such an environment, but its common occurrence fossilized in large numbers suggests mass mortality occurred when water mixing led to suddenly increased salinities. There is a general lack of benthic organisms and of evidence of bioturbation in the laminated Plattenkalk layers, while the presence of a benthic cyanobacterial mat, suggested by micro-ripples on bedding planes, also suggests that grazers were absent. The plants, insects, feathers and occasional tetrapod carcasses must have drifted into the lake in rivers or some may have blown in, so that the majority of the fauna and flora of the Crato Lagerstätte should be considered allochthonous. This is not dissimilar to the situation described in Chapter 10 for the Upper Jurassic Solnhofen Plattenkalks, although the Crato Formation lacks any obvious evidence or mechanism for rapid burial.

196 Fish 'mine' in Santana Formation, near Jardim (see 195).

195 Near Jardim, Chapada do Araripe, Ceará, north-east Brazil.

197 Discarded fossil fish debris at fish 'mine' near Jardim (see 196).

198 'Fisherman' with fossil fish from Santana Formation, Cancau, near Santana do Cariri, Chapada do Araripe, Ceará, north-east Brazil.

199 Manual working of the laminated Crato Formation limestones near Nova Olinda, Chapada do Araripe, Ceará, north-east Brazil.

200 Stone Yard, Nova Olinda, cutting ornamental paving slabs from Crato Formation limestone.

Soft-tissue preservation within the Santana Lagerstätte concretions is very different and is possibly unique in the fossil record. Martill (1988, 1989) argued that the exquisite preservation of delicate tissues such as gills, muscles, stomachs and eggs (**201**) is the result not just of rapid burial, but of rapid fossilization, perhaps instantaneous fossilization.

By the time of deposition of the Santana Formation the saline lakes had dried up and a period of evaporite deposition, represented by the Ipubi Formation, was followed by a marine incursion. Huge numbers of fish migrated into the basin from the open sea and initially survived the resulting brackish waters before a further mass mortality event. Martill (1988, 1989) suggested that the bottom waters were hypersaline and that an upward migration of the halocline could result in mass death, as could a sudden increase in temperature or an algal bloom. The fish are preserved within calcium carbonate concretions, but notably their soft tissue is preserved in calcium phosphate (cryptocrystalline francolite). Martill (1988, 1989) showed that francolite precipitation is enhanced in oxygen deficient, low-pH (acidic) environments as would be caused by the onset of mass decomposition.

By scanning electron microscopy of the phosphatized tissue Martill (1988, 1989) recognized banded muscle fibres with cell nuclei, gill filaments with secondary lamellae, stomach walls and ovaries with eggs. He observed that the gill tissues of fresh trout became bacterially infested and began breaking down within 5 hours of death and had completely disappeared within one week. He therefore concluded that the phosphatization of the Santana fish must have begun within one hour of the fishes' death and he termed this the 'Medusa Effect'. The phosphate was produced by bacteria feeding on proteins in the carcass so that some decay was actually necessary to initiate fossilization.

The next stage in this remarkable preservation is the rapid nucleation after burial of a carbonate concretion around the phosphatized fish to preserve its three-dimensionality. Formation of concretions is known to occur around organic remains in oxygen deficient, low-pH conditions in which lime normally remains in solution, but it requires a local increase in pH in the microenvironment around the decaying carcass (perhaps caused by the release of ammonia?) to allow the lime to precipitate. Both Martill (1988) and Maisey (1991) stressed the contribution to this process made by a putative cyanobacterial mat or scum on the sea floor, as has also been suggested for Ediacara (Chapter 1), Grès à Voltzia (Chapter 7), the Holzmaden Shale (Chapter 8) and the Solnhofen Limestone (Chapter 10).

The same mechanism of fluctuating pH and successive precipitation of calcium phosphate and calcium carbonate has also preserved the segmented limbs in ostracods, the delicate wing membranes of pterosaurs, and comparable structures in other reptiles.

Description of the Santana and Crato Formation biotas
Crato

Insects and other arthropods. This represents the largest and most diverse Cretaceous insect assemblage in the world and includes aquatic, semi-aquatic and terrestrial groups (see Grimaldi, 1990). It is essentially modern at family level and includes mayflies, damselflies and dragonflies (and their larvae), cockroaches and termites, locusts, crickets and grasshoppers, earwigs, leaf-hoppers, true bugs and water bugs, lacewings and snakeflies, beetles, weevils, caddis flies, true flies, wasps and bees. (**202–205**). Below the family level it includes many new and still undescribed forms. Other terrestrial arthropods include scorpions (**206**) and whip scorpions, spiders (**207**), a single japygid dipluran, solifuges and centipedes, again many of them new and undescribed (see Wilson and Martill, 2001 for bibliography). Rare decapod crustaceans complete the arthropod fauna.

Plants. Gymnosperms include common fragments and leafy shoots of cheirolepidiaceous conifers. More important are the numerous early angiosperms (**208**), which include flowers, seeds, fruiting bodies, fruits, leaves and roots. These have only recently begun to be described (e.g. Mohr and Friis, 2000) and many new taxa await description.

201 Close-up of **215**, showing soft muscle tissue in *Notelops brama* (PC).

202 A long-horned grasshopper (PC). Length 60 mm (2.5 in) including antennae.

203 A leafhopper (PC). Length 20 mm (0.8 in).

204 A dragonfly (SCM). Wingspan 150 mm (6 in).

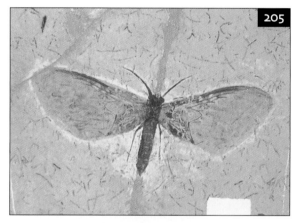

205 A moth (SCM). Wingspan approx. 190 mm (7.5 in).

206 A scorpion (MM). Total length 21 mm (0.8 in).

207 A diplurid spider (HMB). Body length (including spinnerets) 15 mm (0.6 in).

208 The angiosperm *Trifurcatia flabellata* (HMB). Height 115 mm (4.5 in).

Fish. Small specimens of the gonorhynchiform fish *Dastilbe elongatus* are very common (**209**), often with several on a single bedding plane (Davis and Martill, 1999). The gonorhynchiforms, which include the present-day milkfish (*Chanos chanos*), are a sister-group to the carps, minnows and catfish.

Pterosaurs. In recent years the first pterosaurs have been discovered from the Crato Formation, some with soft tissue preservation. Frey and Martill (1994) described an ornithocheirid, *Arthurdactylus*, with a wing-span of 4 m (13 ft), while Campos and Kellner (1997) discovered a remarkable tapejarid complete with its soft tissue head crest (**210**).

Others. Less common, but nonetheless important, are occurrences of a juvenile frog (Maisey, 1991) and occasional feathers (Maisey, 1991; Martill and Filgueira, 1994). Lizards, turtles (with soft tissue) and even a fossil bird (with feathers) have apparently been found recently, but await description (Martill, 2001).

Santana

Fish. The well-preserved fossil fish, often in fish-shaped concretions, from the Santana Formation, have been available from commercial dealers around the world for many years. They occur in vast numbers and include over 20 different taxa, the most common being actinopterygians (ray-finned fish). *Vinctifer* is a long, pike-shaped fish, easily recognized by its deep flank scales and its extended rostrum (**211**, **212**). It is often found with its back arched, suggesting dehydration of body tissues on death. *Tharrhias*, a gonorhynchiform fish related to *Dastilbe* (Crato Formation), often occurs as several individuals within a single concretion (**213**). Other common genera include *Rhacolepis*, which has a thin, fusiform body and a very pointed snout (**214**), and the closely related *Notelops* (**201**, **215**). The distinctive pycnodontid fish *Proscinetes* is a member of the more primitive holostean ray-finned fish (**216**).

Sarcopterygians (lobe-finned fish) are represented by two coelacanths, *Mawsonia* and *Axelrodichthys*, amongst the largest fish known from Santana (over 3 m [10 ft]), and chondrichthyans (cartilaginous fish) are represented by the hybodont shark *Tribodus* recognized by its two dorsal fins with spines, and the ray, *Rhinobatos*.

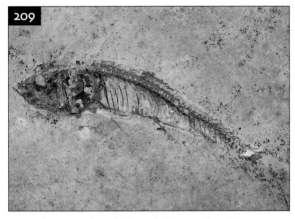

209 The gonorhynchiform fish *Dastilbe elongatus* (PC). Length 40 mm (1.6 in).

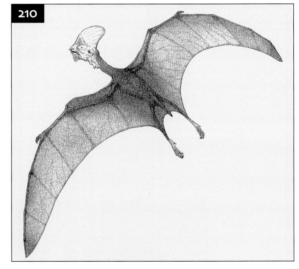

210 Reconstruction of tapejarid pterosaur.

211 The teleostan fish *Vinctifer comptoni*, with elongated flank scales and extended rostrum (PC). Original length 600 mm (24 in).

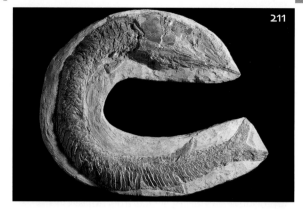

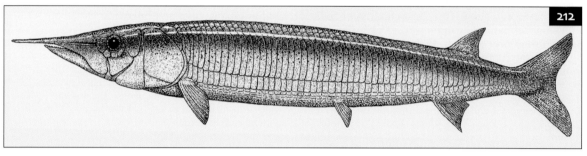

212 Reconstruction of *Vinctifer*.

213 The teleostean fish *Tharrhias araripis* (PC). Length of nodule 780 mm (2.5 ft).

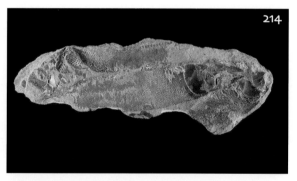

214 The teleostean fish *Rhacolepis buccalis* (CFM). Length of nodule 205 mm (8 in).

215 The teleostean fish *Notelops brama,* showing preservation of soft muscle tissue (PC). Length of nodule 330 mm (13 in).

216 The pycnodontid fish *Proscinetes* sp. (PC). Length 400 mm (16 in).

Pterosaurs. In recent years several well-preserved pterosaurs have been described, some with wing membranes (e.g. Martill and Unwin, 1989). They are all pterodactyloids, and are mostly large forms with wingspans over 5 m (16 ft), many with bizarre skull crests (**217**). There is some disagreement as to whether these forms (*Santanadactylus*, *Araripesaurus*, *Cearadactylus*, *Brasileodactylus*, *Anhanguera*) are related to the pterosaurs from the English Greensand or whether they represent new groups.

Dinosaurs. Recently, a number of theropod dinosaurs have also been described. These include partial skulls of two spinosaurids (*Angaturama*; Kellner, 1996, and *Irritator*; Martill *et al.*, 1996, the latter being a large fish-eating dinosaur with an unusual head crest; **218**), the synsacrum of a possible oviraptosaur (Frey and Martill, 1995), and two small coelurosaurs. The first of these (*Santanaraptor*) is known from the ischia, hindlimbs and caudal vertebrae and includes patches of fossilized skin and muscle fibres (Kellner, 1996, 1999). The second is known from the pelvic girdle, sacrum and partial hindlimbs of a possible compsognathid, and exhibits soft tissue preservation of the intestinal tract and a postpubic air sac (Martill *et al.*, 2000).

Other reptiles. Other tetrapods include two genera of pelomedusid turtles (the oldest known examples of pleurodires or side-necked turtles) and two crocodilians (a terrestrial notosuchid and an aquatic tremato-champsid). The notosuchid genus, *Araripesuchus*, is also known in West Africa and testifies to the existence of a continental link with South America after the origin of this lineage.

Invertebrates. Apart from ostracods, invertebrates are not common, but do include small shrimps, gastropods and bivalves. Two irregular echinoids have been described, but these are rare. There are no ammonoids, belemnoids, nautiloids nor corals, crinoids or brachiopods.

Palaeoecology of the Santana and Crato Formations

The Crato Formation Plattenkalks represent a shallow, stagnant freshwater lake community in which increasing salinity inhibited any autochthonous life except for rare crustaceans and the small fish, *Dastilbe*, which lived in freshwater tongues around deltas. The lake margins supported an abundance of thick vegetation, amongst which lived a variety of aquatic, semi-aquatic and terrestrial insects and spiders, which would have been a rich source of food for lizards, frogs and scorpions. These must all have drifted or been blown into the lake and are allochthonous elements of the biota, as are the rare birds and pterosaurs which flew over the lake.

217 Reconstruction of *Anhanguera*.

Interpretation of the Santana Formation fish fauna is less straightforward; it has been variously described as a fully freshwater, fully marine, and as a marine fauna living in estuaries. Martill (1988) cited the occurrence of echinoderms as evidence for marine conditions, but Maisey (1991) showed that these were from a higher horizon than the fish-bearing concretions, suggesting that fully marine conditions did not occur until later. The lack of normal marine faunas (such as ammonoids, corals, brachiopods) supports this view.

Most Santana fish species are endemic, and although most of the known families are normally marine, almost all of them do have some freshwater members. These contradictory data have led many authors to describe it as a 'quasi-marine' or perhaps brackish water community. Further interpretive research is required.

There is evidence (Maisey, 1991) for at least three distinct fossil assemblages in the Santana Formation, each characteristic of a particular concretion lithology representing three separate collecting sites after which they are named. 'Santana' concretions are usually oval in shape, of small size, and do not reflect the fossil outline. *Tharrhias* is abundant, *Brannerion*, *Araripelepidotes*, and *Calamopleurus* are common, and *Cladocyclus*, *Axelrodichthys*, *Vinctifer* and *Rhinobatos* are rare. Crocodiles, turtles, pterosaurs and plants are also found and the environment is interpreted as clear, oxygenated, near-shore waters.

'Jardim' concretions are large and platy with their shape reflecting the outline of the fossil (**216**). Large specimens of *Rhacolepis* and *Vinctifer* are abundant, *Brannerion*, *Araripelepidotes*, *Cladocyclus*, *Calamopleurus*, *Axelrodichthys* and *Rhinobatos* are common, and *Mawsonia* and *Tharrhias* are rare. Turtles and pterosaurs are rare and the environment is interpreted as a muddy/sandy, anoxic bottom, further from the shore.

'Old Mission' concretions (from Missão Velha) are also large, but are thick rather than platy and do not reflect the fossil outline. *Rhacolepis* and *Vinctifer* are abundant, *Brannerion* is common, and *Araripichthys*, *Calamopleurus* and *Cladocyclus* are rare. Terrestrial and aquatic reptiles are unknown and the environment was deeper, open water with a muddy anoxic bottom.

All three assemblages are dominated by pelagic organisms, mainly fish, but are notable for their absence of pelagic marine invertebrates, such as cephalopods. Moreover, while semi-aquatic reptiles (crocodiles and turtles) do occur, fully marine reptiles such as ichthyosaurs are absent. Epibenthic molluscs (bivalves and gastropods) are known, as are rare echinoids, but other epibenthic marine invertebrates such as corals, crinoids and brachiopods are not and Maisey (1991) suggests that the benthic invertebrates were washed in accidentally. He concludes that the environment was a shallow embayment passing into a coastal region with periodic marine incursions causing mixing of waters.

Comparison of the Santana and Crato Formations with other Cretaceous biotas
Sierra de Montsech, Catalonia, Spain

Although several Plattenkalk facies are known from the Cretaceous, many with soft part preservation, the Santana and Crato biotas are quite distinct and have not been found elsewhere. Some Cretaceous biotas are, however, comparable and none more so than the Sierra de Montsech Lagerstätte exposed on the southern flanks of the Pyrenees in the Province of Lleida in Catalonia, Spain (Martínez-Delclòs, 1991).

Many of the Montsech species, as in Santana/Crato, were endemic but, at higher taxonomic levels, the two biotas do have similarities. The slightly older (Berriasian/Valanginian) Montsech fauna is dominated by fish (at least 16 genera), most of which were brackish water actinopterygians, with one coelacanth (sarcopterygian) and two sharks (chondrichthyans). There were also rare frogs and crocodiles but, significantly, no turtles, pterosaurs or dinosaurs. Bird feathers are known (with one fragmentary skeleton), plus a large insect fauna with some spiders, ostracod and decapod crustaceans, bivalves and gastropods, and well preserved plants, including questionable angiosperms.

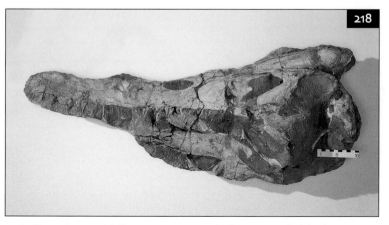

218 The spinosaurid dinosaur *Irritator challengeri*, with faked snout. (SMNS). Estimated true length of skull 800 mm (30 in).

Lithologically, the Plattenkalk is a finely laminated limestone, but it is not homogeneous; wackestones, packstones and bioclastic breccias suggest some degree of depositional slope and slumping. The deposit is interpreted as lacustrine with most of the fauna being autochthonous, living in the depositional basin; trace fossils and coprolites are common at some horizons. However, the thick sequence of undisturbed laminated Plattenkalks and the lack of benthos suggests that deeper parts of the lake had anoxic bottom conditions and were less equitable.

Further Reading

Agassiz, L. 1841. On the fossil fishes found by Mr. Gardner in the Province of Ceará, in the north of Brazil. *Edinburgh New Philosophical Journal* **30**, 82–84.

Beurlen, K. 1962. A geologia da Chapada do Araripe. *Anais da Academia Brasileira de Ciências* **34**, 365–370.

Campos, D. and Kellner, A. W. A. 1997. Short note on the first occurrence of Tapejaridae in the Crato Member (Aptian), Santana Formation, Araripe Basin, Northeast Brazil. *Anais da Academia Brasileira de Ciências* **69**, 83–87.

Davis, S. P. and Martill, D. M. 1999. The gonorynchiform fish *Dastilbe* from the Lower Cretaceous of Brazil. *Palaeontology* **42**, 715–740.

Frey, E. and Martill, D. M. 1994. A new pterosaur from the Crato Formation (Lower Cretaceous, Aptian) of Brazil. *Neues Jahrbuch für Geologie und Paläontologie, Abhandlungen* **1994**, 379–412.

Frey, E. and Martill, D. M. 1995. A possible oviraptosaurid theropod from the Santana Formation (Lower Cretaceous, ?Albian) of Brazil. *Neues Jahrbuch für Geologie und Paläontologie, Monatshefte* **1995**, 397–412.

Gardner, G. 1841. Geological notes made during a journey from the coast into the interior of the Province of Ceará, in the north of Brazil, embracing an account of a deposit of fossil fishes. *Edinburgh New Philosophical Journal* **30**, 75–82.

Gardner, G. 1846. *Travels in the interior of Brazil, principally through the northern provinces.* Reeve, Benham and Reeve, London, xvi + 562 pp. (Reprinted AMS Press, New York, 1970.)

Grimaldi, D. A. 1990. Insects from the Santana Formation, Lower Cretaceous of Brazil. *Bulletin of the American Museum of Natural History* **195**, 1–191.

Jordan, D. S. and Branner, J. C. 1908. The Cretaceous fishes of Ceará, Brazil. *Smithsonian Miscellaneous Collections* **25**, 1–29.

Kellner, A. W. A. 1996. Remarks on Brazilian dinosaurs. *Memoirs of the Queensland Museum* **39**, 611–626.

Kellner, A. W. A. 1999. Short note on a new dinosaur (Theropoda, Coelurosauria) from the Santana Formation (Romualdo Member, Albian), Northeastern Brazil. *Boletim do Museu Nacional, Geologia* **49**, 1–8.

Maisey, J. G. 1991. *Santana fossils: an illustrated atlas.* T. F. H. Publications, New Jersey, 459 pp.

Martill, D. M. 1988. Preservation of fish in the Cretaceous Santana Formation of Brazil. *Palaeontology* **31**, 1–18.

Martill, D. M. 1989. The Medusa effect: instantaneous fossilization. *Geology Today* **5**, 201–205.

Martill, D. M. 1993. *Fossils of the Santana and Crato Formations, Brazil.* (Field Guide to Fossils No.5). The Palaeontological Association, London, 159 pp.

Martill, D. M. 2001. The trade in Brazilian fossils: one palaeontologist's perspective. *The Geological Curator* **7**, 211–218.

Martill, D. M., Cruickshank, A. R. I., Frey, E., Small, P. G. and Clarke, M. 1996. A new crested maniraptoran dinosaur from the Santana Formation (Lower Cretaceous) of Brazil. *Journal of the Geological Society of London* **153**, 5–8.

Martill, D. M. and Filgueira, J. B. M. 1994. A new feather from the Lower Cretaceous of Brazil. *Palaeontology* **37**, 483–487.

Martill, D. M. and Frey, E. 1995. Colour patterning preserved in Lower Cretaceous birds and insects: the Crato Formation of N. E. Brazil. *Neues Jahrbuch für Geologie und Paläontologie, Monatshefte* **1995**, 118–128.

Martill, D. M., Frey, E., Sues, H. D. and Cruickshank, A. R. I. 2000. Skeletal remains of a small theropod dinosaur with associated soft structures from the Lower Cretaceous Santana Formation of northeastern Brazil. *Canadian Journal of Earth Sciences* **37**, 891–900.

Martill, D. M. and Unwin, D. M. 1989. Exceptionally well preserved pterosaur wing membrane from the Cretaceous of Brazil. *Nature* **340**, 138–140.

Martínez-Delclòs, X. 1991. *Les calcàries litogràfiques del Cretaci inferior del Montsec. Deu anys de campanyes paleontològiques.* Institut d'Estudis Ilerdencs, 162 + 106 pp.

Mohr, B. A. R. and Friis, E. M. 2000. Early angiosperms from the Lower Cretaceous Crato Formation (Brazil), a preliminary report. *International Journal of Plant Science* **161**, 155–167.

Small, H. 1913. Geologia e suprimento de água subterrânea no Ceará e parte do Piaui. *Inspectorat Obras contra Secas, Series Geologia* **25**, 1–180.

Spix, J. B. von and Martius, C. F. P. 1823–1831. *Reise in Brasilien.* Munich, 1388pp.

Wilson, H. M. and Martill, D. M. 2001. A new japygid dipluran from the Lower Cretaceous of Brazil. *Palaeontology* **44**, 1025–1031.

GRUBE MESSEL

Chapter Twelve

Background: the Cenozoic Era

The end of the Cretaceous Period, which was also the end of the Mesozoic Era, was marked by a mass extinction event which saw the end of the dinosaurs and pterosaurs on land, and ammonites and marine reptiles in the sea. In the succeeding Cenozoic Era, which consists of the Paleogene, Neogene (together these are commonly referred to as the Tertiary) and Quaternary periods, animals and plants assumed a more modern appearance. Mammals and birds replaced dinosaurs and pterosaurs as the dominant vertebrates on land. After the Cretaceous Period, the great many ecological niches left empty by the extinction of the dinosaurs quickly became filled by mammals and birds in an adaptive radiation. By the middle of the Eocene Epoch (the middle epoch of the Paleogene Period), nearly all of the orders of mammals and major groups of birds had evolved, and there were also present some mammal groups which have since become extinct. At this time, there were no high Alps, nor North Atlantic, and Europe and North America were still connected by land bridges in the vicinity of the Faeroes. There were marine basins in the North Sea, northern France, the Low Countries and Denmark area, and a complex of islands and basins over much of the rest of Europe. Volcanism was common, some associated with the opening of the Atlantic Ocean, and also along old fracture zones such as the Rhine Rift Valley (graben). Grube Messel (which translates the 'Messel Pit' – it was a brown coal open pit until its closure and subsequent conservation as a World Heritage Site in 1995) was situated where the Rhine Graben cut across the Central European Island, and in this down-warping of the Earth's crust an extensive series of lakes was formed. Whether the Grube Messel area represents just one basin within a more extensive river system, part of a much larger lake, or a crater lake, is still debated.

The mammals are the best-known fossils from Grube Messel. Surprisingly, perhaps, most of the mammals at Grube Messel appear to have originated outside Europe and migrated across the region in Eocene times. Mesozoic mammals are rare and primitive, and few Palaeocene fossils exist. The few finds of pre-Eocene mammals in Europe show that they were insectivore-like remnants of Mesozoic types rather than the more modern forms found at Messel. A few mammal types at Messel seem to represent these primitive European forms: the insectivore-like mammals, early hedgehog relatives, and early ungulates. Invasions from elsewhere brought in modern mammals, for example rodents, ant-eaters, horses, bats and primates. The Messel birds belong to modern orders and, despite Messel being a lake deposit, there are almost no water-birds, most being forest-dwellers such as owls, swifts, rollers and woodpeckers. Plant fossils indicate a subtropical climate, with representatives of palms and citrus, for example, but without specialist tropical forest families. Land arthropods do not generally preserve easily, and though rare at Messel, when they do occur they commonly show beautiful colours and patterns. The beetles, in particular, show structural colours (iridescence). Many fish, amphibians (e.g. frogs) and reptiles (e.g. crocodiles and turtles) occur at Messel, which indicates that there was life either in the lake or in tributaries close by.

History of discovery of the Grube Messel biota

The first pit at Messel was dug in 1859 for iron ore, but in 1875 brown coal (actually an oil shale) was discovered and mining of this deposit began. Later that year the first fossil was found – the remains of a crocodile. Over the next century, a great many fossils were discovered, and after major mining activities ceased in the 1960s, more methodical collecting and preparation techniques began. In the early 1970s, the Hessen government planned to use the Messel Pit as a repository for waste material after oil shale extraction had ceased. Immediately, objections were raised by scientists and amateur palaeontologists. By this time, some Messel fossils were attracting high prices and many fossil dealers were collecting at the site, so the pit was closed to the public for safety reasons. Because of the threat of infilling the pit with refuse, a number of German palaeontological institutions began rescue excavations. As a result of these investigations the site's importance became much better known to the palaeontological world. In the late 1970s, special exhibitions on the Messel biota were put on in the Senckenberg Museum, Frankfurt-am-Main, and in the Hessen Museum in Darmstadt. In the early 1980s, permission was granted for refuse disposal to begin at the site. In April 1987, with the threat of complete loss of the fossil site imminent, an international symposium on the Messel Pit was organized at the Senckenberg Museum, but it was not until the early 1990s that the plan to fill Grube Messel with refuse was finally abandoned. The site was eventually recognized as being of international importance to science as a World Heritage Site in 1995, thus preserving the pit for future generations of palaeontologists to continue to excavate and study.

The main collecting technique at Messel involves simply splitting the thin shale slabs with large blades. Once a fossil is discovered, however, special techniques are required to protect the find while transferring it to the laboratory for further study. If the shale is allowed to dry out then the fossil disintegrates, so the most urgent treatment is to keep the shale block wet. The fossil is then transferred to the laboratory while still damp and wrapped in watertight plastic. Once in the laboratory, the transfer technique developed by Kühne (1961) is most useful for vertebrate fossils. Still keeping the fossil damp, the shale is carefully removed from around the bones with needles under a dissecting microscope. As much of the fossil is exposed as possible and then the bones are coated with resin to hold them in place. Now, the whole slab can be turned over and the other side of the fossil exposed in the same way. Eventually, the fossil is completely free from the rock and encased in resin. Insects and plants have to be left in the shale and kept wet both during preparation and afterwards, when glycerine, which is less volatile than water, is used to maintain 'dampness'. A further technique used in the study of Messel vertebrates is the use of X-rays to visualize details of the skull and fine bones (e.g. bat wing bones) which are otherwise difficult to see or uncover from the matrix without damage.

Stratigraphic setting and taphonomy of the Grube Messel biota

From the composition of the mammal fauna at Grube Messel, the strata can be correlated with the lowermost mammal faunas of the Geiseltal lignite beds, which have themselves been correlated with the marine Lutetian Stage (lower Eocene) of the Paris Basin. The Messel Formation represents the lowermost part of the Lutetian Stage, and is therefore about 49 million years in age.

The Messel Formation can be traced only a few kilometres away from Grube Messel, because it was laid down in restricted lake basins within the Rhine Rift Valley (**219**). The lakes were formed in fault-bounded basins and because the faults were active and there was associated volcanic activity, the preserved sediments reflect this. The Messel Formation consists of a base of gravel and sand, which represents an initial high-energy fill resulting from screes at the edge of the basin and/or river activity. The oil shale overlies the sand and gravel, and represents quieter conditions of sedimentation once the lake had formed. Occasional slumps of coarse material from the lake edge, caused by renewed fault activity, can be seen as lenses of sand and gravel within the oil shale sequence. Slumps of shale within the shale sequence can also be seen. With continued fault movement, the lake basin was constantly being rejuvenated, and its shape and extent were probably also changing; the thickest sequence of oil shale recorded is 190 m (620 ft).

The present-day oil shale consists of clay minerals (i.e. the 'shale'), organic matter (15%, kerogen – the 'oil'), and some 40% water. This makes the Messel oil shale a valuable source of oil for exploitation by man. Under normal circumstances, organic matter would decay in water by the action of bacteria, through the process of oxidation, but at Messel in the Eocene something prevented this from happening. The environment suggested for Messel at that time was subtropical forest, and a great deal of vegetable matter, especially algal remains, would have accumulated on the lake bed. If the bottom waters of the lake lacked oxygen (anoxia), then the organic matter would decay only partly or not at all, resulting in the accumulation of kerogen and, more importantly for palaeontologists, the preservation of the soft parts of animals and plants. It is possible that the sheer volume of algae blooming in the lake at times resulted in an excess of their decaying remains, which used up all of the available oxygen and rendered the bottom waters anoxic.

Description of the Grube Messel biota

Plants. As today, the angiosperms (flowering plants) dominate the flora at Grube Messel, but there is some evidence of other plant groups. Fronds found at Messel belong to the modern marsh and mangrove fern families Osmundaceae (Royal ferns), Schizaceae and Polypodiaceae. Gymnosperms include the families Cephalotaxaceae (Chinese plum-yews), Cupressaceae, Taxodiaceae and Pinaceae. The Taxodiaceae are the swamp-cypresses which grow in permanently inundated conditions around the Gulf of Mexico and as far north as Virginia. Pinaceae (Pines), on the other hand, are generally found in drier conditions. The relative scarcity of gymnosperms in the Messel lake deposits suggests that they did not live very close to the lake, and the floral composition provides evidence of a subtropical to warm-temperate climate.

Among angiosperms, true grasses (Graminae) had not, by mid-Eocene times, evolved the diversity we see today. However, grass-like plants such as rushes (Juncaceae), sedges (Cyperaceae) and reed-maces (Typhaceae) were present; such plants prefer wetter conditions than grasses. An interesting family, which shows a Gondwanan (mainly southern hemisphere) distribution today in wet habitats, is the Restionaceae. Pollen of this family occurs at Messel, as well as world-wide in late Cretaceous and early Tertiary deposits. It appears that the Restionaceae were displaced by the radiation of the Graminae in mid-Tertiary times. Other monocotyledons present at Grube Messel include the palms (Palmae, Arecaceae), which are typical of tropical and subtropical climates today as well as in Eocene times, and arums and lilies, which typically prefer damp conditions.

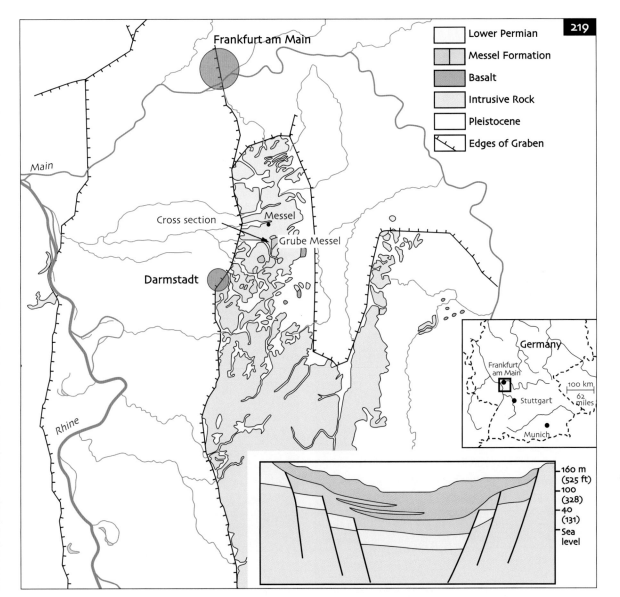

219 Locality map and cross section of Grube Messel (after Schaal and Ziegler, 1992).

An important family of woody, evergreen trees and shrubs in Tertiary times was the Lauraceae, the laurels. Again, their presence indicates subtropical conditions. Another climatic indicator family present at Grube Messel is the Menispermaceae (moon-seed); these plants occur mainly as lianas in tropical and subtropical forests today. Other plant families of mostly tropical–subtropical climates, remains of which have been found commonly at Grube Messel, are the Theaceae (tea), Icacinaceae (a small family of vines), Vitaceae (grape), Rutaceae (citrus) and Juglandaceae (walnut). One of the commonest leaf fossils found at Grube Messel is that of the water-lily family, Nymphaeaceae (**220**). The presence of this family cannot give us climatic information but it suggests that shallow, open, oxygenated water conditions were present near to the site of deposition of the oil shale. Mention should also be made of a very important plant family which occurs commonly at Messel: the Leguminosae. This family has an almost worldwide distribution today and was clearly common in the Eocene too.

Judging from the abundant leaf, fruit, seed, spore and pollen remains found at Grube Messel, the forests surrounding the lake must have been diverse and lush in a subtropical climate. However, the lake has clearly sampled a number of different habitats: open water, swamp, bank-side, damp forest and drier regions. For example, a family of plants restricted today to south-east Asia but common in other Tertiary floras is the Mastixiaceae. However, this family is rare at Messel, and it has been suggested that they did not occur close to the lake so their large fruits were unable to be carried into the site of deposition. Similarly, Fagaceae (beeches, chestnuts and oaks) is a widespread family mainly outside the tropics today. Distinctive remains occur commonly at other fossil sites but at Messel only their pollen has been found, suggesting that these trees occupied higher, drier ground further away from the lake.

There is abundant evidence at Messel of plant interactions with other organisms. For example, leaves occur with characteristic fungal (rust) spots, evidence of insect eggs on leaves and larval chewing marks. Plant remains have been found in the gastrointestinal tracts of some mammals; for example, grape seeds have been found inside the small horse *Propalaeotherium*, together with leaves of laurel, walnut and other families, indicating a varied diet for this animal. Pollen has been found under the elytra (wing cases) of beetles, which implicates them as pollinators.

Arthropods. Some of the most beautiful of Messel fossils are the beetles, with well-preserved structural coloration (iridescence) on their elytra (**221**). Because of their relatively tough bodies, beetles (Coleoptera) are the most common (63%) of all insect remains found at Messel. Amongst the Coleoptera in the

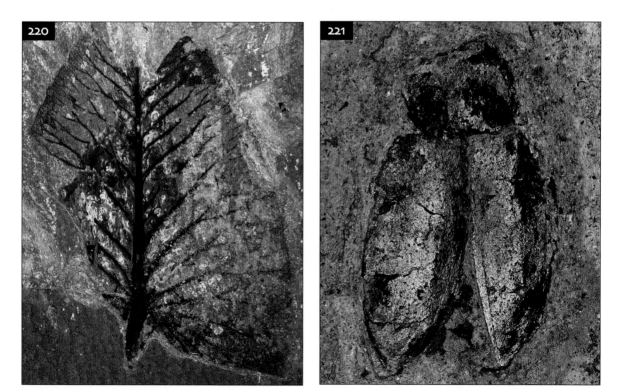

220 Leaf of water-lily (Nymphaeaceae) (SMFM). Length 160 mm (c. 6 in).

221 Iridescence on the elytra (wing cases) of a leaf beetle (Chrysomelidae) (SMFM). Length 4.5 mm (0.2 in).

collection at the Senckenberg Museum, Frankfurt, click beetles (Elateridae) are the most common (15.8%), followed by weevils (Curculionidae, 12.8%), jewel beetles (Buprestidae, 8.4%), dung beetles (Scarabaeidae, 3.9%), stag beetles (Lucanidae, 1.7%), ground beetles (Carabidae, 1.4%), water beetles (Dascillidae, 1.4%), longhorn beetles (Cerambycidae, 0.5%), and rove beetles (Staphylinidae, 0.26%). Other families are present in smaller percentages and include the colourful leaf beetles (Chrysomelidae) and the ground beetle family Tenebrionidae. Larvae of the water-beetle genus *Eubrianax*, which can only survive in highly oxygenated water, such as at waterfalls, were a surprise find at Grube Messel, and indicate that these animals, as well as many others, perhaps, were washed in from elsewhere.

Hymenoptera (ants, bees and wasps) are the second-most common (17%) insects from Messel. Of these, ants (Formicidae) are the most frequent, and these are almost entirely known from winged stages. Of especial interest are fossils of giant queens with wingspans up to 160 mm (c. 6 in) across. At this size they exceed all known Hymenoptera, let alone ants. Other Hymenoptera from Messel include the parasitic wasps (Ichneumonidae), Chalcidae, Tiphiidae, Scoliidae, potter wasps (Eumeniidae), Anthophoridae, spider-hunting wasps (Pompilidae) and digger wasps (Sphecidae). Interestingly, no plant wasps (Symphyta) have been found, and this may be explained by their abundance today in temperate, rather than tropical, climates.

Bugs (Heteroptera) form 12.5% of the insect fauna at Messel, and are mostly represented by the family Cydnidae (> 80%); these are generally plant-sucking bugs, but have been observed in the tropics to favour juices from animal carcasses which were, of course, abundant around the Messel lake. Other ground- and plant-dwelling insects occurring at Messel include cockroaches (Blattodea, 1.5%) and crickets (Orthoptera, 0.5%). Diptera (flies, 0.4%) and Lepidoptera (butterflies and moths, 0.25%) are strangely rare at Messel compared to other Tertiary insect localities until it is realized that, with their large wings, these insects would have floated on the water surface rather than sunk to the lake floor, and so it is not surprising that very few dragonflies (Odonata), stoneflies (Plecoptera) and caddisfly adults (Trichoptera) have been found at Grube Messel either.

Arachnids (spiders, harvestmen, mites, ticks, scorpions and their allies) are rare as fossils outside amber (see Chapter 13), even in sediments which preserve good insect fossils. A handful of specimens occur at Grube Messel, and most seem to belong to the orb-weaver spiders. This is not surprising since such species are common in lake-side vegetation today. A single specimen of a harvestman (Opiliones) has been found at Grube Messel.

Fish. All fish preserved at Grube Messel belong to the advanced bony fishes (osteichthyans: neopterygians) but show a wide diversity. One of the most frequently encountered Messel fish is *Atractosteus*, a gar. The gars are distinctive predators with large heads and massive jaws; their scales are of a large, overlapping, ganoid type with shiny enamel, resembling armour plating. The bowfins are represented by *Cyclurus*, which

is the most frequently encountered fish at Messel. Both bowfins and gars are about 200–300 mm (8–12 in) in length, although smaller and larger (up to 500 mm [20 in]) specimens do occur. Like gars, bowfins are also formidable predators. A single specimen of an eel, *Anguilla ignota*, about 600 mm (2 ft) long, was a striking find. Classified in the modern genus of catadromous eels (i.e. those which are born in the sea and return to it to spawn but spend much of their lives in fresh water), its presence suggests that the Messel lake had a connection to the sea; but why, when in modern fresh waters where eels occur they are abundant, has only one specimen ever been found at Messel?

Amphibians. Lakes are usually thought of as excellent places in which to find amphibians – tetrapod vertebrates which return to water to breed even if they spend much of their time on land. Yet at Grube Messel only a single specimen of salamander, *Chelotriton*, has been found, and very few anurans (frogs and toads). Nevertheless, the preservation of the frogs is excellent (**222**); some show skin and muscles preserved, and there are tadpoles too.

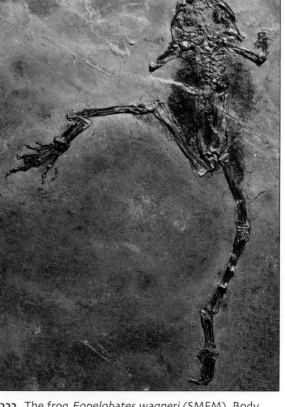

222 The frog *Eopelobates wagneri* (SMFM). Body length (excluding legs) about 6 cm (c. 2.4 in).

Reptiles. Freshwater turtles generally make good fossils with their bony shells. Completely removed from the oil shale, the skeleton of *Trionyx* makes a beautiful specimen (**223**). Trionychid turtles today are almost entirely aquatic, leaving the water only to lay eggs. Crocodiles are also associated with water and, not surprisingly, many specimens have been found at Grube Messel. Some complete skeletons have been recovered, from juveniles up to 4 m (13 ft) long adults. In all, six genera occur at Messel, which is a high diversity for a single site compared to the present day. It has been suggested that only one genus, *Diplocynodon*, actually lived in the lake, because a full set of juvenile stages of this genus, the commonest, have been found. All the others are presumed to have been washed in from adjacent rivers or other parts of the complex lake system.

Much rarer than the semi-aquatic crocodiles and turtles at Messel are the terrestrial lizards and snakes. A variety of lizards has been found at Messel, from large, predatory monitors, through agile iguanids, to limbless forms. Snakes are relative latecomers amongst reptiles, the first fossils occurring in Cretaceous rocks. By the Eocene, the familiar constrictors *Boa* and *Python* were present and have been found at Messel as the genus *Palaeopython* (**224**), while the more advanced venomous snakes came later and none occur at Messel.

Birds. A great variety of birds can be found at Grube Messel. Palaeognathous birds today include only the ratites (the flightless ostriches, rheas and allies) and the cursorial tinamous of South America. Grube Messel yields

223 The turtle *Trionyx* (SMFM). Carapace length 300 mm (1 ft).

224 The beautifully preserved snake *Palaeopython* (SMFM). Length about 2 m (6 ft).

Palaeotis, an ostrich-like palaeognathan which could represent an ancestor to modern ratites. All other birds are classified as Neognathae, the sister-group to Palaeognathae. There is some evidence for birds of prey and chicken-like birds. The primitive ibis *Rhynchaeitis messelensis* was the first bird found at Messel, in 1898. Giant, flightless birds dominated the scene in early Tertiary times. A femur imprint of a diatryma (a giant with a powerful beak) has been found at Messel, together with three specimens of the large *Aenigmavis*. Other birds in the same group as these, the Gruiformes (crane-like birds),

are found at Messel, including seriemas (South American stork-like birds) and rails. Among Charadriformes (plover-like birds) there is a flamingo preserved at Messel: *Juncitarsus*. The Messel owl, *Palaeoglaux*, preserves a strange plumage unknown in modern owls; possibly it was not yet nocturnal. There are representatives of the nightjars, swifts, and rollers (**225**). The last is a group of birds with colourful plumage: kingfishers, hoopoes, hornbills and bee-eaters as well as rollers. Some of the Messel rollers are preserved with exceptionally fine plumage (but without colour, unfortunately) and very falcon-like grasping claws. A number of birds at Messel can be assigned to the woodpeckers, which provides further evidence of forest nearby. Indeed, the single specimen of a flamingo is the only evidence of a true water-bird; all of the others are terrestrial or forest-dwelling forms.

Mammals. Grube Messel is best known for its fossil mammals. Many groups are present, including marsupials in the form of opossums. At least two forms are present: a small, tree-climbing form with a prehensile tail, and a larger, shorter-tailed genus which was probably a ground-dweller. Placentals form the bulk of the mammal fauna at Messel. A fascinating creature called *Leptictidium* (**226, 227**) was originally pigeon-holed with the insectivores, but is now included with the Proteutheria: primitive placentals from the Cretaceous and early Tertiary.

225 A roller-like bird displaying fine preservation of feathers (SMFM). Total length about 200 mm (8 in).

226 The unusual mammal *Leptictidium nasutum* (SMFM). Total length 750 mm (30 in) (including tail which is 450 mm [18 in]).

227 Reconstruction of *Leptictidium*.

Three species of *Leptictidium* have been described from Grube Messel, and what makes these animals striking is their locomotion. They had relatively huge hind-legs, small fore-legs, and an exceptionally long tail consisting of more than 40 vertebrae (this is far more than in any known mammal today). The tail was not prehensile, so we can only assume that it acted as a balancing organ and the animal used its large hind-legs for locomotion. However, unlike kangaroos and jerboas, which use their hind-legs to jump along, the weak joints of the hind-legs of *Leptictidium* suggest that it loped along with a running gait unknown in extant mammals. Also in Proteutheria is *Buxolestes*, which has short, strong feet and a thick tail; its resemblance to otters suggests that it was a good swimmer, and gut contents of fish bones and scales confirm this. Among true insectivores are highly specialized hedgehog relatives, including *Macrocranion* with large hind-legs adapted to jumping, and *Pholidocercus*, which had a scaly tail and a large, strongly innervated organ on its forehead. Another insectivore, *Heterohyus*, was arboreal and had long second and third digits on its fore-limbs for extracting insect grubs from tree-holes.

Bats are rarely preserved in sedimentary deposits because of their aerial locomotion and cryptic habits, yet at Grube Messel they are the most common mammal fossils. The suggested reason for this is that they were overcome by noxious gases emanating from the Messel lake and drowned. The hundreds of bats preserved at Messel form a unique resource for the study of bat evolution. All are Microchiroptera (insectivorous, rather than fruit-eating, bats). The excellent preservation in the oil shale includes skin and wing membrane, muscle, fur, and gut contents (**228**). The different wing aspect ratios of the six species described suggest a diversity of habits including high fliers among the trees; lower-level, open-space hunters; and foliage or ground-level skimmers. The gut contents include lepidopteran (moth) scales, trichopteran (caddisfly) hairs and coleopteran (beetle) wing-cases. All used echo-location for hunting. Three-quarters of the specimens belong to the group of ground-level fliers, which would be more susceptible to the noxious gases than the high-fliers.

Four finds of primates from Messel belong to the Adapiformes; animals half the size of a cat, they are possible relatives of the Madagascan lemurs. Pangolins (order Pholidota) are nocturnal ant-eaters with body scales instead of hair. When disturbed they roll into a tight ball, protected by their scaly covering. Many of their features are primitive and suggest an early origin, perhaps in the Cretaceous Period. The discovery of the earliest and best-preserved fossil pangolins at Messel, which are almost identical to modern pangolins, is consistent with this hypothesis. However, in spite of what seemed obvious adaptations for ant- and termite-eating (myrmecophagy), the stomach contents of the fossils consisted almost entirely of plant fragments. One intriguing explanation for this paradox is that these early pangolins were actually feeding on pieces of leaf stolen from leaf-cutter ants, common denizens of tropical forests, and that ant-eating evolved from this habit. The other order of mammals alive today which specializes in myrmecophagy is the Edentata, of which a single specimen has been found at Messel. Its gut contents included insect cuticle together with sand grains (helpful in grinding up the insects) and woody tissues, suggesting it had been eating wood termites.

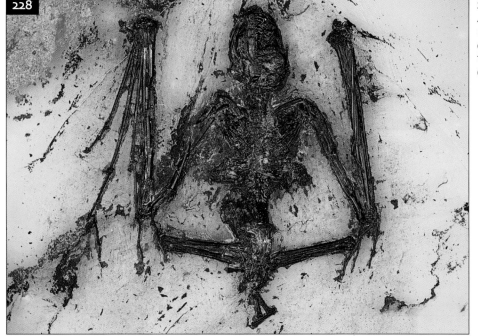

228 The bat *Archaeonycteris trigonodon* (SMFM). Length of forearm 52.5 mm (2 in).

Rodents are characterized by their large, single pair of incisors, an adaptation for nibbling vegetation and gnawing seeds. Three genera are known from Messel: a large, squirrel-like rodent with leaves in its gut, and two smaller, mouse-like forms. The rodents would have formed part of the diet of carnivores such as the possibly arboreal *Paroodectes*, a miacid (an extinct family seemingly ancestral to modern Carnivora), and the hyaenodontid creodont *Proviverra*. Hoofed mammals were present at Messel, both early representatives of modern groups and extinct, ancestral forms. An example of the latter is *Kopidodon* which, though many aspects of its morphology indicate that it belongs among the ancestors of hoofed mammals, has strong claws and a long tail consistent with an arboreal lifestyle. The odd-toed ungulates (Perissodactyla) include horses, tapirs and rhinos, and Messel has provided some of the finest fossil examples of these groups. The Messel horse *Propalaeotherium* (**229**), is known from more than 70 specimens ranging from foals to full-grown adults, and there are two species, one with an adult shoulder-height of 300–350 mm (12–14 in) and one 550–600 mm (22–24 in) – small horses indeed! Other features suggest they were primitive horses; for example, the fore-legs had four hooves and the hind-legs three (in modern horses each leg has only a single hoof). The gut contents of these early horses indicate that they ate leaves and seeds (browsers); grasslands and grazing horses had not yet developed in the Eocene. The even-toed ungulates (Artiodactyla) include cattle, deer, pigs, camels, giraffe and hippopotamus. Two genera of primitive artiodactyls occur at Grube Messel which, like the perissodactyls, were dog-sized browsers or foragers on forest litter.

Palaeoecology of the Grube Messel biota

While the cause of anoxia (see p. 122) seems clear, the physiographic setting of the Messel lake is still hotly debated. One theory is that there was an extensive river system across the area, under which the Rhine Graben began to form, thus isolating parts of the river system as lakes in depressions in the hilly terrain. Evidence for this theory comes from the distribution and orientation of fish and caddisfly fossils, which are both inhabitants of flowing water. They indicate that water flowed into, and presumably out of, the Messel lake. Another proposal, based on studies which showed that not only was the lake in existence for some 100,000 years but the sediments across the exposed area are remarkably uniform, is that Messel lake was a large, persistent basin and that the fossil-bearing regions represent deeper, anoxic depressions within this much larger structure. In this model, many of the fossils were inhabitants of the larger lake which had oxygenated waters. A third model explains the Messel lake as a crater lake within a maar volcanic depression. Maar volcanoes are explosive eruptions of ash which result in circular depressions, commonly later filled with lakes. Such maar lakes dating from Tertiary times occur throughout southern Germany, e.g. Randecker Maar. In this model, the lake would have been deep and steep-sided – ideal conditions for bottom-water anoxia. In addition, toxic gases emitted from the lake could have downed over-flying bats and birds, while terrestrial mammals could easily have slipped down the steep banks. Local streams could have been home to caddis larvae and small fish, but lacked larger fish and water-birds, which are rare as

229 The Messel horse *Propalaeotherium parvulum* (SMFM). Height at shoulder 300–350 mm (12–14 in).

fossils. Similarly, while wind-blown leaves and flowers occur, twigs and branches of trees are rare. At present, it is difficult to choose between these three models and more research is necessary to evaluate them.

Throughout this chapter, reference has been made to what the fossil finds tell palaeontologists about the ecology of the Messel area. The evidence comes from comparison with the habits of their modern relatives, sedimentological setting, and consideration of the taphonomy (preservational history) of the fossils. The picture painted by the evidence is a complex but fascinating one. Messel appears not to preserve a typical lake biota, but to have sampled widely from the surrounding forests as well as different parts of the lake. It is undoubtedly the forest-dwelling forms which have attracted most interest among palaeontologists, because such a habitat is rarely preserved in the fossil record. In this respect, comparison could be made with amber biotas (see Chapter 13).

Comparison of Grube Messel with other Tertiary biotas

The brown coal seams of Geiseltal, some 150 km (90 miles) to the north-east of Messel, have been excavated for three centuries, and over that time have yielded a vast quantity of fossils which can be compared to those found at Grube Messel (Voigt, 1988). The vegetation, for example, is similar in many ways, but there are important differences which can be explained because Messel was a lake which sampled vegetation from a wide variety of habitats; the Geiseltal lignite represents a bog habitat, an ecosystem which has a characteristically impoverished biodiversity. There are interesting differences in the fauna between the sites too. More than 300 specimens of an axolotl-like amphibian have been found at Geiseltal, but none at Messel, though the reptilian faunas of the two sites are comparable. Because of its unique setting, bat fossils are relatively common at Messel, but not at Geiseltal; yet other mammals are more comparable – for example, primates, ant-eaters and ungulates.

Further Reading

Buffetaut, E. 1988. The ziphodont mesosuchian crocodile from Messel: a reassessment. *Courier Forschungsinstitut Senckenberg* **107**, 211–221.

Franzen, J. L. 1985. Exceptional preservation of Eocene vertebrates in the lake deposits of Grube Messel (West Germany). *Philosophical Transactions of the Royal Society of London*, Series B **311**, 181–186.

Habersetzer, J. and Storch, G. 1990. Ecology and echolocation of the Eocene Messel bats. 213–233. *In* Hanak, V., Horacek, T. and Gaisler, J. (eds.). *European bat research 1987*. Charles University Press, Prague.

Habersetzer, J., Richter, G. and Storch, G. 1992. Palaeoecology of the Middle Eocene Messel bats. *Historical Biology* **8**, 235–260.

Kühne, W. G. 1961. Präparation von flachen Wirbeltierfossilien auf künstlicher Matrix. *Paläontologische Zeitschrift* **35**, 251–252.

Lutz, H. 1987. Die Insekten-Thanatocoenose aus dem Mittel-Eozän der 'Grube Messel' bei Darmstadt: Erste Ergebnisse. *Courier Forschungsinstitut Senckenberg* **91**, 189–201.

Maier, W., Richter, G. and Storch, G. 1986. *Lepticidium nasutum* – ein archaisches Säugetier aus Messel mit aussergewöhnlichen biologischen Anpassungen. *Natur und Museum* **116**, 1–19.

Novacek, M. 1985. Evidence for echolocation in the oldest known bats. *Nature* **315**, 140–141.

Peters, D. S. 1989. Ein vollständiges Exemplar von *Palaeotis weigelti. Courier Forschungsinstitut Senckenberg* **107**, 223–233.

Schaal, S. and Ziegler, W. (eds.). 1992. *Messel. An insight into the history of life and of the Earth.* Clarendon Press, Oxford, 322 pp.

Sturm, M. 1978. Maw contents of an Eocene horse (Propalaeotherium) out of the oil shale of Messel near Darmstadt. *Courier Forschungsinstitut Senckenberg* **30**, 120–122.

Voigt, E. 1988. Preservation of soft tissues in the Eocene lignite of the Geiseltal near Halle. 325–343. *In* Franzen, J. L. and Michaelis, W. (eds.). *The Eocene at Lake Messel. An International Symposium.* Courier Forschungsinstitut Senckenberg **107**. Senckenbergische Naturforschende Gesellschaft, Frankfurt, 452 pp.

Westphal, F. 1980. *Chelotriton robustus*, n. sp., ein Salamandride aus dem Eozän der Grube Messel bei Darmstadt. *Senckenbergiana Lethaea* **60**, 475–487.

BALTIC AMBER

Chapter Thirteen

Background: forest life in the Cenozoic Era

In the last chapter we saw how the Messel lake sampled the plants and animals from a wide variety of habitats both near and far from the lake site. While the larger mammals and birds are the most important fossils from Messel, more delicate flying insects, such as flies (Diptera) and butterflies (Lepidoptera), are rare because they tended to float on the lake surface rather than sink to the bottom. However, the tendency of these insects to adhere to the water surface is what makes them far more likely to be preserved in amber – fossilized tree resin – to which delicate insects and other animals are attracted and by which they become engulfed. Tree resin is a very localized deposit and, while some trees produce copious amounts of exudate, it is most likely to preserve animals and plants which are associated with trees. Rapid removal of a carcass from a decaying environment is the best way of preserving it, and what could be quicker than trapping and engulfing an insect in seemingly impermeable resin? There is no initial transport of the carcass, apart from some flow down the tree trunk and struggling by the animal itself, although later transport of the amber is usually necessary for its concentration into a sedimentary deposit.

Amber samples forest life from different sources than a lake deposit, and its method of preservation is far better than that of lacustrine sediments for delicate invertebrates such as insects, while vertebrates are rarely preserved in amber. Insects living on tree bark are the most likely to be found in amber. Many insects are attracted to tree resin, possibly sensing the volatile oils given off by the exudate, but whether this attraction benefits the tree or the insects is not known. Once attracted to, or overcome by, the resin, insects become trapped in it owing to its adhesiveness. Predators such as spiders are attracted to the struggling insects and then they, too, become trapped, in a similar manner to the predators preserved in the tar pits of Rancho La Brea (Chapter 14). Animals living in bark crevices, in moss on or at the foot of a tree, flying insects in the amber forest, and their predators are the most common animals preserved in amber, and a wide range of plant

material, such as spores, pollen, seeds, leaves and hairs, is also commonly found embedded in the amber. Small drops of resin are unlikely to collect many organisms, but some trees exude vast quantities of resin from wounds; the large masses of resin produced from these cracks, termed 'Schlauben' (Schlüter, 1990), flow down the tree trunk and form ideal traps.

Resin is produced by a variety of trees today, and in the past. A prolific producer of resin is the araucarian (monkey-puzzle family) Kauri pine, *Agathis australis* (**230**), which grows in northern New Zealand. Amber deposits from this tree are known in New Zealand from some 40 Ma in age. Younger deposits of copal (resin which is not as well fossilized as amber), 30,000–40,000 years old, also occur in New Zealand and formed the basis of a copal mining industry in the last century. Copal will melt with low heat and it was used in the past to make varnish and moulded into objects such as trinkets and

230 Young Kauri pine, *Agathis australis*, a prolific producer of resin since Tertiary times, North Island, New Zealand.

even false teeth. During the amberization process, fresh resin first loses its volatile oils, then polymerization begins. Once the resin has hardened (to 1–2 on the Mohs Scale) and is no longer pliable it is called copal. However, in this state it will dissolve in some organic solvents and melts at a low temperature (below 150°C). Copal (especially African) does contain insect and other inclusions but, being much younger than amber, is generally of less interest to the palaeontologist. To form true amber, polymerization and oxidation must continue for a longer period of time, until the material has reached a hardness of 2–3 on the Mohs Scale, will not melt below about 200°C, and is not soluble in organic solvents. These physical properties of amber and copal are useful to remember because genuine amber with inclusions can command a good price in the gem trade, so fakes, often simply copal or made with melted copal, can be recognized. One excellent example of a forgery was discovered by Andrew Ross of the Natural History Museum in London (Grimaldi *et al.*, 1994). He was interested in a fly in apparently genuine amber which belonged to an advanced modern family otherwise unknown in the fossil record. On examination under the microscope with a rather warm lamp, a crack appeared. Further inspection revealed that a piece of amber had been cut in half, one side hollowed out and the (modern) fly inserted, then glued carefully back together again!

History of discovery of Baltic amber

Amber was familiar and of special significance to ancient civilizations; it has been found in jewellery dating to before 10,000 BC. Amber was called *succinum* (sap-stone) by the Romans and *elektron* by the ancient Greeks; the English word *electricity* is derived from the static electric effect produced when amber is rubbed by a soft cloth. Pliny, in the first century AD, was the first person to describe the properties of amber and correctly determine its origin as the fossilized resin from trees; other ideas of its origin concerned the tears of deities or dried excretions of beasts. Pliny recognized that the traded amber originated from the north of Europe, and so Baltic amber is both the oldest recorded and best known of all amber deposits.

Owing to its beauty (**231**), and hence value, trade in amber has been an integral part of Baltic cultures, and from ancient times to the present day, amber has been collected from the shores of the Baltic Sea. The Teutonic Knights, who occupied the Baltic Sea coast in the thirteenth to fifteenth centuries, appropriated control of the amber trade from the Prussians. After the Teutonic Knights were defeated at the battle of Grunwald (Tannenberg) in 1410 by the combined armies of Poland and Lithuania, their monopoly over the amber market crumbled, and other powers struggled for its control. In the middle of the nineteenth century, an enterprise emerged to dredge amber from the sea and mine it from the ground. A factory was established near the village of Palmnicken (now Jantarny) on the west coast of Samland region of Russia. (Samland is the old name for a peninsula in the Russian enclave called Kaliningrad Oblast – capital Kaliningrad – squeezed between Poland and Lithuania.) Between

231 Baltic amber with inclusions (PC). Length about 70 mm (2.8 in).

1875 and 1914 the factory produced between 225,000 and 500,000 tons of raw amber per year. The best quality was used for jewellery and sculptures but the inferior grades were melted down for varnish. The finest amber carving ever produced was the Amber Room, commissioned in 1701 by King Frederick I of Prussia. This consisted of wall panels in variously coloured amber pieces depicting scenes. It took ten years to complete, and was then transported to Peter the Great's summer palace in St Petersburg. However, during the Second World War the room was removed to Königsberg Castle, on the Baltic coast of (at that time) Germany. Later, when the Russians were advancing on Königsberg (Kaliningrad), the room panels were again dismantled and apparently packed into crates in the castle dungeons. What happened to the Amber Room after that remains a mystery; at least one person amongst those searching for it since the war is known to have lost his life in mysterious circumstances.

Awareness of amber inclusions for their biological interest is also very old. The finest collection ever amassed was that of the Königsberg University Geological Institute Museum in Samland. This collection was started when dredging and mining began in 1860. Originally thought to have been destroyed by bombing during the Second World War, the collection was actually dispersed to other museums. Huge collections of Baltic amber can now be found in, for example, the Natural History Museum in London, the Museum of the Earth in Warsaw, the Zoological Institute in St Petersburg, and the Humboldt University Museum in Berlin.

Stratigraphic setting and taphonomy of Baltic amber

Figure **232** shows the presumed location of the Baltic amber forest, the coastal outcrop in Samland, and the places where reworked amber can be collected along the shoreline. Amber lumps washed out of the forest in streams were incorporated into a glauconitic clay called the Blue Earth. Glauconite is an iron mineral with a typically green colour but which in a dark clay could appear bluish. A number of layers of Blue Earth occur in a sequence of sandstones and mudrocks (**233**) which, by their marine fossil content, have been dated to mainly

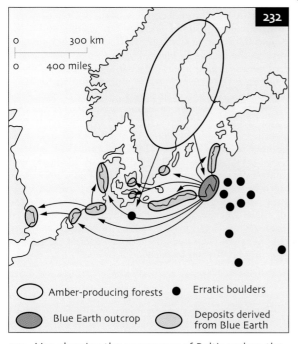

232 Map showing the occurrence of Baltic amber: the presumed location of the amber forest, outcrop of the Samland Blue Earth deposits, and deposits derived from the Blue Earth outcrop (after Schlüter, 1990).

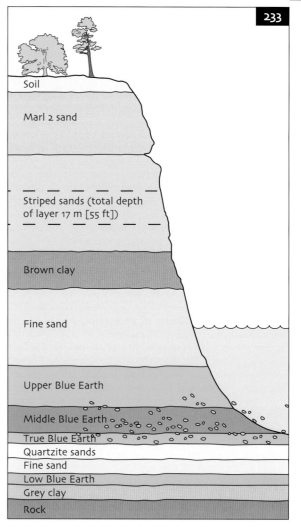

233 Stratigraphic log of the Samland Blue Earth showing the location of amber-rich layers (after Poinar, 1992).

between the mid-Eocene and early Oligocene Epochs (Paleogene Period). It is the Blue Earth which is quarried for amber near Palmnicken. During the Pleistocene Epoch, ice sheets scoured the area now occupied by the Baltic Sea, and Blue Earth became incorporated into the glacial boulder clay (till) which now covers much of northern Poland, Germany and Denmark. Because of its low density, amber from outcrops of Blue Earth or boulder clay derived from the deposit is easily washed out by currents and carried to the beach. Thus, Baltic amber can be found on the coastline of not only the Baltic Sea but also the North Sea.

There has been much discussion about the type of tree which produced the resin which became the Baltic amber. The first studies in the 1830s placed it as a species of the extant pine genus *Pinus*, but later anatomical investigations placed it in an extinct genus and species: *Pinites succinifera*. Other anatomical studies in the last century suggested that the wood was closer to the spruce genus *Picea*. The more modern technique of infra-red spectroscopy, which is useful for distinguishing different types of amber, reveals a characteristic flattish region to the resulting spectroscopic curve for Baltic amber (called the 'Baltic shoulder'). Care must be taken when comparing spectroscopic curves for ambers with those for resins because the spectrum varies with degrees of fossilization. Nevertheless, comparisons have been made which reveal that the infra-red spectrum of Baltic amber is closer to that of the New Zealand araucarian *Agathis australis*. More recent analyses using pyrolysis gas spectrometry have produced similar results. So, the

spectrum of Baltic amber points to Araucariaceae, morphology to Pinaceae. Furthermore, while *Agathis* is a copious producer of resin today and *Pinus* produces far lesser amounts, there is fossil evidence of *Pinus* in Baltic amber but none of araucarians. A possible compromise, suggested by Larsson (1978), is that the tree belonged to an extinct group of gymnosperms with characteristics of both Pinaceae and Araucariaceae (it could not be ancestral to these families because they were both present in the Cretaceous Period).

At first sight, insects in amber appear to be preserved in three dimensions, with their original cuticle and coloration, but as empty husks lacking any internal organs. This appearance was first shown to be misleading as long ago as 1903, when Kornilovich described striated muscles in Baltic amber insects. Later, Petrunkevitch (1950) recognized internal organs in Baltic amber spiders, and the scanning electron microscope studies of Mierzejewski (1976a,b) showed that the preservation of the internal structure of spiders in Baltic amber (including spinning glands, book-lungs, liver, muscles and haemolymph cells) was better than that shown by the insects. Using the transmission electron microscope, Poinar and Hess (1982) revealed muscle fibres, cell nuclei, ribosomes, endoplasmic reticulum and mitochondria in a Baltic amber gnat. Organisms are preserved in amber by a process known as mummification. In this process, dehydration of the tissues results in their shrinkage to some 30% of their original volume (thus giving the fossil the appearance of an empty husk). The organic material is not removed from oxidation because amber does allow slow gaseous diffusion (so cannot be used in the study of ancient atmospheres as had been hoped) but amber has fixative and anti-bacterial properties, like many resins. The ancient Egyptians used resins in the preparation of their mummies, and the anti-bacterial properties of many resins is well known; the distinctive flavour of Greek *retsina* wine is due to the use of resin to prevent it going off.

Because of the excellent preservation of structures at the subcellular level there has been considerable interest in the possibility of recovering pieces of the macromolecule deoxyribonucleic acid (DNA) from amber-preserved organisms. In the movie *Jurassic Park* it was suggested that if fossilized dinosaur blood could be extracted from the gut of a gnat preserved in Mesozoic amber, the DNA sequence of the dinosaur thus revealed could then be used to generate living dinosaurs. Like all the best science fiction, this idea is based on possibility. However, while attempts have been made to extract DNA from ambers, none has been successful, and since it is now known that amber is not as impermeable as was once thought, it is highly unlikely that the molecule could have survived for millions of years (or even for more than a few hours after death – it is known that DNA breaks down rapidly after cell death) without degradation.

Even if an inclusion is well preserved, there may be cracks, cloudiness or other inclusions which obscure important features of the organism. Sometimes the inclusion occurs at the edge of the amber piece, so only part of it is preserved. A common feature of Baltic amber is a whitish fuzz, resembling mould, surrounding the inclusion (**234**). Under the microscope, this material – called 'emulsion' by Petrunkevitch (1942) – is seen to consist of tiny air bubbles (Mierzejewski, 1978). It is believed that the emulsion is caused by moisture escaping from the carcass and reacting with the resin early in the taphonomic process. Though it is especially characteristic of Baltic amber, it has been noted in other ambers such as the early Cretaceous amber of the Isle of Wight (Selden, 2002), and its presence or absence may be related to differences in water solubility of the resins from different amber-producing trees (Poinar, 1992).

Description of the Baltic amber biota

Since the first scientific studies of Baltic amber some 250 years ago, a huge variety of organisms have been reported as inclusions. The bulk of Poinar's (1992)

book is devoted to descriptions of the variety of plants and animals reported from amber inclusions, and Baltic amber figures in nearly every section. For more detailed reviews the reader is referred to this work, to the books by Larsson (1978) and Weitschat and Wichard (2002), which are restricted to Baltic amber, and to Keilbach (1982) which covers arthropods.

Some 750 species of plants have been described from Baltic amber over the past two-and-a-half centuries, principally between 1830 and 1937, and were summarized by Conwentz (1886), Göppert and Berendt (1845), Göppert and Menge (1883), and Caspary and Klebs (1907). The main studies on the Baltic amber tree were by Conwentz (1890) and Schubert (1961). However, according to a study by Czeczott (1961), only 216 of these 750 are valid species, including 5 species of bacteria, 1 myxomycete (slime mould), 18 fungi, 2 lichens, 18 liverworts, 17 mosses, 2 ferns, 52 gymnosperms and 101 angiosperms.

234 Wedge-shaped beetle (Coleoptera: Rhipiphoridae) in Baltic amber (GPMH). Note the covering of white emulsion. Length about 5 mm (0.2 in).

235 Rose in Baltic amber (Angiospermae: Rosaceae) (GPMH). About 5 mm (0.2 in) across.

Fungi. Most of the fungi known from Baltic amber are saprophytic growths on other organisms. Lichens are a symbiotic intergrowth of algae and fungi, and are commonly found on tree trunks. Bryophytes (mosses and liverworts) are generally rare in the fossil record, and some of the best preserved bryophytes occur in Baltic amber, described by Czeczott (1961). Pteridophytes (ferns) are rarely preserved in amber, but Czeczott (1961) reported two species: *Pecopteris humboldtiana* and *Alethopteris serrata*.

Gymnosperms. The gymnosperms in Baltic amber mainly belong to living genera, and it is interesting that the most closely related living species to the Baltic amber species occur in North America, east Asia and Africa. The genera represented in Baltic amber include the cycad *Zamiphyllum*; the podocarp *Podocarpites*; Pinaceae *Pinus*, *Piceites*, *Larix* and *Abies*; Taxodiaceae *Glyptostrobus*, *Sciadopitys* and *Sequoia*; and the Cupressaceae *Widdringtonites*, *Thujites*, *Librocedrus*, *Chamaecyparis*, *Cupressites*, *Cupressinanthus* and *Juniperus*.

Angiosperms. About two-thirds of the angiosperms (flowering plants) in Baltic amber have been identified from flowers, fruits or seeds (**235**, **236**); the remainder were described from leaves and twigs. Most belong to extant genera, and a great diversity of families is represented, belonging to temperate, Mediterranean, subtropical, and even tropical groups. As with the gymnosperms, the distribution of closely related living species of angiosperms is interesting. For example, the fossil genus *Drimysophyllum* (Magnoliaceae) most closely resembles the extant *Drimys* of the Winteraceae, which now occurs in the Malay Archipelago, New Caledonia, New Zealand and Central and South America. The nearest present-day locality for *Clethra*, found in Baltic amber, is Madeira. Other tropical or subtropical families represented in Baltic amber are: Palmae, Lauraceae, Dilleniaceae, Myrsinaceae, Ternstroemiaceae, Commeinaceae, Araceae and Connaraceae. According to Czeczott (1961) about 23% of the families are tropical, 12% are restricted to temperate regions, and the remainder are cosmopolitan or have a discontinuous distribution. One very common inclusion in Baltic amber, which can be helpful in diagnosing amber from this source, are stellate hairs (trichomes) from oak leaves. Look at the underside of an oak leaf with a hand-lens and these hairs can easily be seen. Oak bud scales are particularly thickly covered by them and, in the spring, oak woods become carpeted with masses of these little star-shaped objects.

Nematodes and molluscs. Microscopic roundworms (Nematoda) are common just about everywhere, so it is not surprising that they have been reported in Baltic amber. Figure **237** shows a parasitic nematode emerging from its host, a midge. Land snails (Mollusca: Gastropoda: Pulmonata and Prosobranchia) are, perhaps surprisingly, rare in Baltic amber, though a few genera have been reported. Klebs (1886) gave a table of where related living species originated, and the areas include central and southern Europe, North America, central Asia, south China and India.

Crustaceans. Arthropods are, of course, the most common inclusions found in amber. Perhaps surprisingly, Crustacea are known from Baltic amber in the form of a few specimens of the amphipod *Palaeogammarus* and the isopods *Ligidium*, *Trichoniscoides*, *Oniscus* and *Porcellio*. Amphipods (sandhoppers and their allies) occur today in moist litter as well as freshwater habitats, and isopods (woodlice) are common inhabitants of litter and damp places.

Myriapods. Myriapods include the predatory centipedes (Chilopoda) and mainly detritivorous millipedes (Diplopoda), both of which are common in terrestrial habitats, especially where it is damp, such as beneath decaying logs and in leaf litter. A wide variety of chilopods have been described from Baltic amber, including the extant genera *Scutigera*, *Cryptops*, *Geophilus*, *Scolopendra* and *Lithobius*. Among millipedes, bristly, soft-cuticled members of the primitive subclass Penicillata, which are common on tree trunks, are represented by three genera. Other millipedes in Baltic

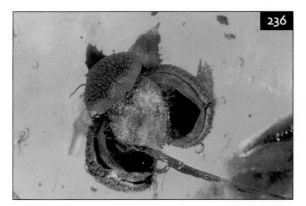

236 Unidentified nut in Baltic amber (GPMH). About 5 mm (0.2 in) across.

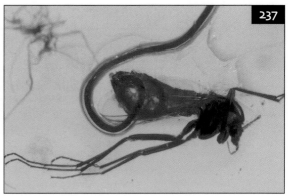

237 Mermithid nematode emerging from the body of its host, a female chironomid midge (GPMH). Length about 6 mm (0.2 in).

amber include the pill millipede *Glomeris*, and other recent genera such as *Craspedosoma*, *Julus* and *Polyzonium*. The specimen shown in figure **238** appears to be a chordeumatid.

Hexapods (insects). Primitive hexapods (insects and related groups) are represented in Baltic amber by two species of dipluran and many Collembola (springtails). The latter group is quite likely to get preserved in amber because of their saltatory locomotion. Silverfish (Thysanura), common today under bark and stones, are represented in the Baltic amber by numerous genera in three families.

More advanced, winged insects are the most common inclusions in amber. A great many mayflies (Ephemeroptera) have been described from Baltic amber. The larvae of these insects are aquatic – indicating that there must have been fresh water in the amber forest – but the adults are short-lived and often swarm in great abundance. Like mayflies, stoneflies (Plecoptera) have aquatic larvae, and representatives of four families occur in Baltic amber. In contrast to mayflies and stoneflies, dragonflies and damselflies (Odonata) are strong fliers, and much less likely to be blown onto sticky resin. It is not surprising, therefore, that they are rare in Baltic amber. Common detritivores, the cockroaches (Blattodea) are relatively common and diverse in Baltic amber. The genera represented are mainly subtropical and tropical forms today. Crickets and grasshoppers, too, occur in Baltic amber, perhaps partly because they are likely to jump onto the resin. Praying mantids and stick insects also have representatives in Baltic amber. Turn over any piece of bark and one insect which is almost certain to be encountered is an earwig (Dermaptera), and therefore it is not surprising to find that they are known from Baltic amber too.

Isoptera (termites) are well-known inhabitants of rotting wood, and they are represented quite commonly in Baltic amber by three families and numerous genera, the commonest species being *Reticulotermes antiquus*. Embioptera are a small group of insects with a thin cuticle and poor powers of flight. They live communally in silken tubes beneath bark and under stones, and one species, *Electrombia antiqua*, is known from Baltic amber. Amongst other small orders of insects with Baltic amber representatives are the Psocoptera (book-lice) (**239**). These little creatures are detritivores on dead plant and animal matter; they are commonly found feeding on the binding paste of old books – hence their common name. In the wild they occur on tree trunks and under bark, so are common in amber. Representatives of eight families and numerous genera are known from Baltic amber. The majority of psocids from Baltic amber are closest to species which today live in tropical and subtropical Asia, America and Africa. Thrips (Thysanura) are also tiny insects, with sucking mouthparts for extracting sap from living vegetation, as well as fluids from other animal and plant tissue; many are abundant pests. Six families have been reported from Baltic amber, and most belong to the family Thripidae.

The order Hemiptera (bugs) is characterized by sucking mouthparts, and can be conveniently divided into two suborders: Homoptera and Heteroptera. Homopterans include cicadas, aphids, leafhoppers and scale insects. Homopterans in Baltic amber include many species of aphids, some scale insects and leafhoppers. In contrast to the abundance of these small homopterans, the large, powerful flying cicadas are rare in Baltic amber. Heteropterans are all the other kinds of bugs, such as shield bugs, water boatmen, water striders, plant bugs, bed-bugs and assassin bugs. They all have piercing mouthparts for feeding on plant sap or animal fluids. Baltic amber includes representatives of nearly all of the groups of Heteroptera. It is understandable that plant-sucking bugs of the large family Miridae, for example, are common in amber, but the records of such creatures as *Nepa* (the large, predatory water-scorpion), water boatmen and water striders are curious. However, these animals do fly regularly, so could be attracted to insect prey trapped in resin.

The order Neuroptera covers a range of related groups: alder-flies, snake-flies and dobson-flies (Megaloptera) and lacewings, ant lions and mantispids

238 Millipede (Diplopoda: Chordeumatida?) in Baltic amber (GPMH). Length about 8 mm (0.3 in).

239 Parasitic mite (Arachnida: Acari: Erythraeidae: *Leptus* sp.) on a book-louse (Psocoptera: Caecilidae) in Baltic amber (GPMH). Body length about 0.5 mm (0.02 in).

(Planipennia). These are large insects, predatory, but generally slow fliers. A few species of most of these groups have been recorded from Baltic amber. Mecoptera (scorpion-flies) have characteristic curved ends to their abdomens, hence their name, but otherwise resemble neuropterans. Four genera have been reported from Baltic amber.

Beetles (Coleoptera) are the most diverse order of insects, and hence the most diverse group of terrestrial animals. Not surprisingly, there are many records of beetles from Baltic amber. Ground beetles (Carabidae) are common on tree trunks and so in amber too. Like the water bugs, water beetles (Dytiscidae, Gyrinidae) have been found in Baltic amber, indicating water bodies in the forested areas. Anobiidae (death-watch beetles) are common wood-boring beetles, and so occur quite commonly in amber. Other wood-boring beetles also occur in amber: the beautiful Buprestidae, the longhorn beetles of the family Cerambycidae, for example. The large family of often brightly coloured leaf beetles (Chrysomelidae) have more than two dozen genera know from Baltic amber. A few genera of ladybirds (Coccinellidae) are known from the amber. A number of families of beetles are associated with bark, especially through their habit of eating bark fungi, and these families are therefore common in amber. Mention should also be made of the superfamily Curculionoidea, the weevils, of which about 50 genera are known from the Baltic amber; and the Elateridae (click beetles), which are very common in the amber (about 40 genera), perhaps because they were attracted to the resin. Scarabaeidae (scarab beetles) are generally associated with animal dung, but a few genera occur in Baltic amber. One of the largest groups of beetles is the family Staphylinidae (rove beetles), and so not surprisingly about 50 genera have been reported from Baltic amber. Figure **234** shows an adult wedge-shaped beetle (Rhipiphoridae), whose modern relatives live on flower heads and whose larvae are parasitic on hymenopterans which visit flowers.

Caddis flies (Trichoptera) have aquatic larvae but their imagines (adults) fly. Ulmer (1912) produced an extensive monograph on the Baltic amber caddises in which 152 species in 56 genera and 12 families were described. This huge diversity (and great number, since they are extremely common in Baltic amber) must reflect a generous supply of rivers and ponds in the amber forest to support the larval stages. Interestingly, while a quarter of all living caddis flies belong to the family Limnephilidae, none from this family occurs in the Baltic amber. The reason for this may be that limnephilids inhabit temperate rather than tropical or subtropical climates. Caddis flies have hairy wings while Lepidoptera (butterflies and moths) have scaly ones. The larger Lepidoptera (macrolepidopterans) are strong fliers and, while many are attracted to resin seeps, are unlikely to become entrapped. Indeed, most representatives of the Lepidoptera found in Baltic amber are microlepidopterans – small moths. Members of the families Eriocraniidae, Nepticulidae, Incurvariidae, Psychidae (bagworms), Tineidae (clothes moths), Lyonetiidae, Gracillariidae, Plutellidae, Yponomeutidae (ermine moths), Elachistidae, Oecophoridae, Scythrididae, Torticidae (tortrix moths), Pyralidae (plume moths) and Adelidae have all been found in Baltic amber. Of these, tineids and oecophorids are the most common, while the large family of tortrix moths is under-represented apparently because they were not easily trapped in the amber. There have been reports of macrolepidopterans captured in Baltic amber, including Sphingidae (hawkmoths), Arctiidae (tiger moths), Noctuidae, Papilionidae (swallowtail butterflies) and Lycaenidae (blues).

Diptera (flies) are the second largest order of insects after the Coleoptera, and are easily recognized by having only a single pair of wings (forewings); the hind wings are reduced to balancing organs called halteres. Three suborders are recognized: Nematocera, Brachycera and Cyclorrhapha. Nematocera are the midges and fungus gnats and among the commonest insects found in amber. Among the many families found in Baltic amber, it is worth mentioning the Anisopodidae (wood-gnats) (**240**), whose larvae live on decaying organic matter; Bibionidae (March flies), the adults of which appear in large numbers in the spring; Cecidomyiidae (gall midges), whose larvae cause leaf galls; Ceratopogonidae (biting midges), whose adults inflict painful bites on animals (including humans) and whose larvae are detritivores in aquatic habitats; Chironomidae (true midges), whose adults swarm but do not bite and whose larvae are aquatic (**237**); Culicidae (mosquitoes), infamous blood-suckers with aquatic larvae; Mycetophilidae and Sciaridae (fungus gnats), tiny flies ubiquitous in woodlands with larvae feeding on fungi and decaying wood (though some mycetophilids prey on insects by attracting them to their glowing tails); Psychodidae (moth flies and sand flies), whose larvae feed on decaying vegetation while the adults are blood-suckers; Scatopsidae (scavenger flies), with larvae that scavenge on rotting organic matter and excrement; Simuliidae (blackflies), more blood-suckers; Tipulidae (crane flies), the familiar 'daddy-long-legs' flies with about 40 genera in Baltic amber; and the Trichoceridae (winter crane flies), which inhabit dark places and are seen mainly in the winter. Baltic amber Brachycera include representatives

240 Adult wood-gnat (Diptera: Anisopodidae) just hatched from its pupal case, Baltic amber (GPMH). Pupa about 4.6 mm (0.2 in) long.

of the following families: Acroceridae (small-headed flies), internal parasites of spiders; Asilidae (robber flies), large, predatory flies; Bombylidae (bee flies), which suck nectar and generally resemble bees; Dolichopodidae (long-legged flies), whose larvae and imagines are both predatory, the former frequently occurring under bark; Empididae (dance flies), whose larvae live under bark (hence their abundance in amber) and whose adults characteristically form little dancing swarms; Rhagionidae (snipe flies), with predatory larvae and imagines (**241**); Stratiomyidae (soldier flies), large, colourful predatory flies with larvae that live under bark; Tabanidae (horse flies), which have familiar, painfully blood-sucking, female adults and aquatic larvae; Therevidae; Xylomyidae; and Xylophagidae. Eighteen families of Cyclorrhapha have been reported from Baltic amber, the best known of which are Drosophilidae (fruit flies) and Muscidae (house flies); each family is represented in Baltic amber by only a single species.

Siphonaptera (fleas) are rare in the fossil record, yet two species are known from the Baltic amber, both belonging to the genus *Palaeopsylla*, which are ecto-parasites of shrews and moles.

Hymenoptera can be subdivided into two suborders: Symphyta (sawflies and horntails) and Apocrita (ants, bees and wasps). Seven families of Symphyta have been recorded from Baltic amber, mainly from single examples of their larvae. There are many different groups of Apocrita: parasitic, solitary, colonial, social, winged and wingless. The parasitic forms (Parasitica) are important controllers of insects numbers, and many are known from Baltic amber, particularly braconids (**242**) and ichneumons. The presence of gall wasps is usually indicated by their galls, such as oak apples, on vegetation; a number of these occur in Baltic amber. Aculeate Apocrita includes the ants, bees and wasps. Ants are abundant in ambers, probably because of their habit of crawling up and down tree trunks. More than 50 genera of ants are known from Baltic amber. It was

because of the types of ants in the Baltic amber – one group associated with a subtropical climate and the other with a temperate one – that Wheeler (1915) suggested the deposition of amber over a long period of time when the climate cooled and the forest type changed. Two species of wasp (Vespidae) have been reported from Baltic amber. Sphecids and pompilids are solitary wasps which dig burrows into which they deposit their eggs alongside narcotized prey (usually spiders) to provide nourishment for the developing larvae when they hatch. A number of sphecids and fewer pompilids have been reported from Baltic amber. Bees (Apoidea) include both the solitary and social forms. A number of species of both types are known from Baltic amber.

Before leaving the insects, mention must be made of an order of orthopteroids discovered as recently as 2002: the Mantophasmatodea (Klass *et al.*, 2002; Zompro *et al.*, 2002) (**243**). First described as a member of an unknown order of insects from Baltic amber in 2001 (Zompro, 2001), extant specimens were later identifed from Tanzania and Namibia. Nocturnal, pedatory insects, their modern habitat (dry, stony mountains) differs considerably from their habitat in the Baltic amber forest.

Arachnids. The fossil record of arachnids (spiders, scorpions, mites and their allies) would be much poorer were it not for Baltic and other amber inclusions. However, most Baltic amber specimens belong to modern families and genera, so tell us little about the evolution of the groups. Scorpions are represented in Baltic amber by six specimens of buthoids, each described in its own genus, and which are most closely related to extant genera from Africa and Asia (Lourenço and Weitschat, 2000). Scorpions are much rarer in Baltic amber than in other ambers. Resembling miniature scorpions but without a tail are the pseudoscorpions. These little animals live in moss, under bark, and in leaf litter, and quite a number of genera in nine families have been found in Baltic

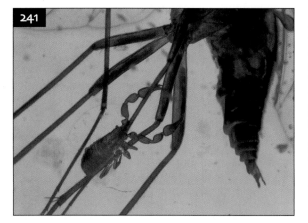

241 Phoresy: pseudoscorpion hitching a ride on the leg of a rhagionid fly, Baltic amber (GPMH). Pseudoscorpion body about 2.5 mm (0.1 in) long.

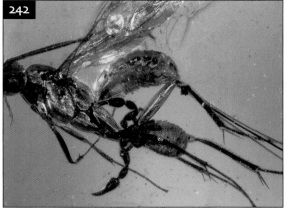

242 Phoresy: pseudoscorpion (*Oligochernes bachofeni*) on a braconid wasp (Hymenoptera: Braconidae) in Baltic amber (GPMH). Body of pseudoscorpion (excluding chelae) about 2.5 mm (0.1 in) long.

amber (Schawaller, 1978). Pseudoscorpions have a clever way of dispersing from one habitat to another. They wait for a flying insect to land on the moss, bark or litter, then clamp onto the leg of the insect with their clawed pedipalps (**241**, **242**). When the insect flies off, it takes the pseudoscorpion with it, and when it lands in a suitable habitat, the pseudoscorpion unhitches. This type of 'hitch-hiking' is called phoresy, and examples of pseudoscorpions phoretic on braconid wasps have been found in Baltic amber (Bachofen-Echt, 1949). Harvestmen (Opiliones) (**244**) are familiar long-legged arachnids which commonly aggregate in huge numbers under loose bark. Nine genera have been described from Baltic amber, but the actual number of specimens is far greater because of the harvestman's habit of shedding a leg if it gets caught, so there are many isolated, unidentifiable harvestman legs in the amber. Mites and ticks are the most diverse animals on land, after the four largest insect orders. Many are parasitic on plants and animals and have sucking mouthparts for

this. Many of the parasitic forms are host-specific, so it is sometimes possible to infer the presence of the host from the fossil evidence of the parasite (**239**). One tick species, *Ixodes succineus*, has been described from Baltic amber, and more than 60 mite genera. Most of the mites belong to the free-living forms which are common detritivores and fungivores today under bark and in moss and litter.

More than 90% of all fossil spiders (Araneae) known are from amber. Representatives of some 33 families have been described from Baltic amber, originally mainly by Koch and Berendt (1854), revised by Petrunkevitch (1942, 1950, 1958), with additions by Wunderlich (1986, 1988). One family, Archaeidae (**245**, **246**), was first described from Baltic amber and only later found living in the Afrotropical and Australian regions. The families are those which would be expected in a forest setting, and many of the genera suggest a subtropical climate. Apart from some small specimens, those living under bark, and moulted skins, it is quite likely that many of the spiders

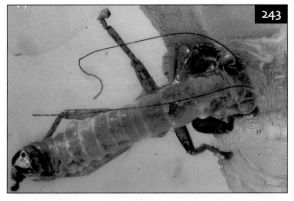

243 *Raptophasma kerneggeri*: type specimen of the recently described order Mantophasmatodea, Baltic amber (GPMH). Body length 11.7 mm (0.5 in).

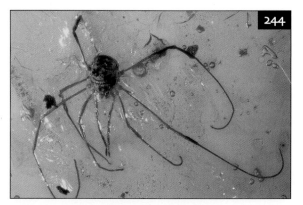

244 Harvestman (Arachnida: Opiliones: *Dicranopalpus* sp.) in Baltic amber (GPMH). Body (excluding legs) about 3.25 mm (0.1 in) long.

245 Male archaeid spider in Baltic amber (GPMH). Length (including chelicerae) 5.625 mm (0.22 in).

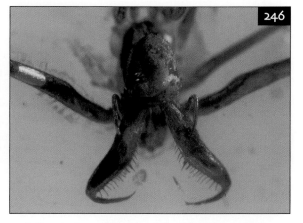

246 Front view of female archaeid spider in Baltic amber (GPMH). Length from back of head to end of chelicerae about 3 mm (0.1 in).

preserved in amber were attempting to prey upon insects already trapped in resin. Amazingly, spider silk and webs (**247**, **248**) have been preserved in Baltic amber, and a pair of spiders were even caught in the act of mating (Wunderlich, 1982).

Vertebrates. These animals are generally too large to be trapped in resin, but some remains have been found. Unfortunately, most reports of frogs and lizards in Baltic amber have turned out, on closer inspection, to be forgeries; or else the specimens have been lost. Bird feathers occur in Baltic amber (e.g. Weitschat, 1980), and they have been referred to sparrows, woodpeckers, nuthatches and tits. Mammals are represented in Baltic amber only by a footprint and some hairs. The hairs have been identified as those of dormice, squirrels and bats.

Palaeoecology of Baltic amber

A number of suggestions have been put forward attempting to explain the assortment of Baltic amber plants that occur today in different climatic zones and widespread parts of the world. Heer (1859) explained the mixture of floral types by proposing that the forest was vast, extending over an area from present-day Scandinavia to Germany and Poland. In this way, the trees could occupy temperate to subtropical climates, possibly extending up the sides of mountains in the northern part of the region. Wheeler (1915) suggested that the Baltic amber forest occupied the same, restricted, geographical area but that the climate changed over the period of amber production. The production of Baltic amber probably extended over several million years, so the forest could have subtly changed from subtropical to temperate (or *vice versa*). Ander (1941) suggested that the forest occupied a humid, mountainous region in which tropical species could live on the southern slopes, while those preferring lower temperatures would be restricted to cooler valleys with a northerly aspect. Abel (1935) and other workers compared the palaeoecology of the Baltic amber forest with present-day southern Florida, where isolated islands (called 'hammocks') of tropical plants occur among subtropical and temperate vegetation. In this region one can find three groups of plants, palms, pines and oaks, occurring together, as in the Baltic amber vegetation. All of these suggestions seem perfectly plausible, but perhaps evidence from other sources is

necessary before we can decide between them. Clearly, Baltic amber sampled the plants and animals of a forest. Climatic evidence indicates that it was temperate to tropical, and the forest was certainly wet because of the number of insects associated with water, particularly as evidenced through their aquatic larvae.

Comparison of Baltic amber with other ambers

Schlüter (1990) compared the Baltic amber fauna with that of the Dominican Republic, the only other amber to have produced sufficient inclusions to make quantitative comparisons with Baltic amber, with informative results. Diptera account for some 50% of inclusions in the Baltic amber, less than 40% in the Dominican. In Dominican amber, Hymenoptera (mainly ants) are second only to Diptera (also nearly 40%) in abundance, but these insects account for only about 5% of the Baltic amber fauna. The reason for this is that ants make up a more disproportionate percentage of the fauna in the tropics than they do in temperate zones, and Dominican Republic amber shows every sign of representing a tropical forest (Poinar and Poinar, 1999). The Dominican Republic amber was produced by a leguminous tree, *Hymenaea protera*, which has as its closest relative *H. verrucosa* of East Africa. It is generally rather paler than Baltic amber and clearer, lacking the abundant oak hairs and the annoying emulsion which often obscures inclusions in Baltic amber. Amber and copal come from a variety of localities in the Dominican Republic, ranging in age from 15 to 45 Ma. Today, Dominican Republic amber is turning up inclusions which equal or, in some cases (e.g. vertebrates), exceed the significance of those of Baltic amber for enlightening us about life in the forests of the Cenozoic Era.

248 Spider web in Baltic amber (GPMH). Picture about 30 mm (1.2 in) long.

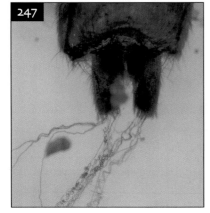

247 Close-up view of spider spinnerets showing silk strands emerging from the spigots, Baltic amber (GPMH). Spinnerets about 0.2 mm (0.008 in) long.

Further Reading

Abel, O. 1935. *Vorzeitliche Lebensspuren.* Gustav Fischer, Jena, xv + 644pp.

Ander, K. 1941. Die Insektenfauna des Baltischen Bernstein nebst damit verknüpften zoogeographischen Problemen. *Lunds Universitets Årsskrift* N. F. **38**, 3–82.

Bachofen-Echt, A. 1949. *Der Bernstein und seine Einschlüsse.* Springer-Verlag, Wien, 204 pp.

Caspary, R. and Klebs, R. 1907. Die Flora des Bernsteins u. andere fossiler Harze des ostpreussichen Tertiärs. *Abhandlungen der Königlich Preussischen Geologischen Landesanstalt. Berlin* N.F. **4**, 1–182.

Conwentz, H. 1886. *Die Flora des Bernsteins. 2. Die Angiospermen des Bernsteins.* Danzig, 140 pp.

Conwentz, H. 1890. *Monographie der baltischen Bernsteinbäume.* Danzig, 203 pp.

Czeczott, H. 1961. The flora of the Baltic amber and its age. *Prace Muzeum Ziemi. Warszawa* **4**, 119–145.

Göppert, H.R. and Berendt, G.C. 1845. *Der Bernstein und die in ihm befindlichen Pflanzenreste der Vorwelt.* Vol. 1. Berlin.

Göppert, H. R. and Menge, A. 1883. *Die Flora des Bernsteins unde ihre Beziehungen zur Flora der tertiarformation und der Oegenwart.* Volume 1. Danzig.

Grimaldi, D. A. 1996. *Amber: window to the past.* Harry N. Abrams Inc. and American Museum of Natural History, New York, 216 pp.

Grimaldi, D. A., Shedrinsky, A., Ross, A. and Baer, N. S. 1994. Forgeries of fossils in 'amber': history, identification and case studies. *Curator* **37**, 251–274.

Heer, O. 1859. *Flora Tertiaria Helvetiae: die tertiäre Flora der Schweiz.* Volume 3. Wintherthur, 377 pp.

Keilbach, R. 1982. Bibliographie und Liste der Arten tierischer Einschlüsse in fossilen Harzen sowie ihrer Aufbewahrungsorte. *Deutsche Entomologische Zeitschrift* N. F. **29**, 129–286, 301–491.

Klass, K.-D., Zompro, O., Kristensen, N. P. and Adis, J. 2002. Mantophasmatodea: a new insect order with extant members in the Afrotropics. *Science* **296**, 1456–9.

Klebs, R. 1886. Gastropoden im Bernstein. *Jahrbuch der Königlich Preussischen Geologischen Landesanstalt und Bergakademie zu Berlin* **188**, 366–394.

Koch C. L. and Berendt, G. C. 1854. Die im Bernstein befindlichen Crustaceen, Myriapoden, Arachniden und Apteren der Vorwelt. *In* Berendt, G. C. (Menge, A., ed.) *Die im Bernstein befindlichen Organischen Reste der Vorwelt.* Berlin, **1** (2), p. 1–124, pl. I-XVIII.

Kornilovich, N. 1903. Has the structure of striated muscle of insects in amber been preserved? *Protokol Obshchestva Estestvoisptatele pri Imperatorskom Yur'evskom Universitete. Yur'ev. (Dorpat)* **13**, 198–206.

Larsson, S. G. 1978. *Baltic amber – a palaeobiological study.* Entomonograph Volume 1. Scandinavian Science Press Ltd., Klampenborg, Denmark. 192 pp.

Lourenço, W. R. and Weitschat, W. 2000. New fossil scorpions from the Baltic amber – implications for Cenozoic biodiversity. *Mitteilungen aus dem Geologisch-Paläontologisches Institut der Universität Hamburg* **84**, 247–59.

Mierzejewski, P. 1976a. Scanning electron microscope studies on the fossilization of Baltic amber spiders (preliminary note). *Annals of the Medical Section of the Polish Academy of Sciences* **21**, 81–2.

Mierzejewski, P. 1976b. On application of scanning electron microscope to the study of organic inclusions from Baltic amber. *Rocznik Polskiego Towarzystwa Geologicznego (w Krakowie)* **46**, 291–5.

Mierzejewski, P. 1978. Electron microscope study on the milky impurities covering arthropod inclusions in Baltic amber. *Prace Muzeum Ziemi. Warszawa* **28**, 79–84.

Petrunkevitch, A. 1942. A study of amber spiders. *Transactions of the Connecticut Academy of Arts and Sciences* **34**, 119–464.

Petrunkevitch, A. 1950. Baltic amber spiders in the collections of the Museum of Comparative Zoology. *Bulletin of the Museum of Comparative Zoology, Harvard University* **103**, 259–337.

Petrunkevitch, A. 1958. Amber spiders in European collections. *Transactions of the Connecticut Academy of Arts and Sciences* **41**, 97–400.

Poinar, G. O. 1992. *Life in amber.* Stanford University Press, Stanford, California, xiii + 350 pp.

Poinar, G. O. and Hess, R. 1982. Ultrastructure of 40-million-year-old insect tissue. *Science* **215**, 1241–2.

Poinar, G. O. and Poinar, R. 1999. *The amber forest: a reconstruction of a vanished world.* Princeton University Press, Princeton, New Jersey, xviii + 239 pp.

Rice, P. C. 1993. *Amber: the golden gem of the ages.* 1993 Revision. The Kosciuszko Foundation Inc., NY, x + 289 pp.

Ross, A. 1998. *Amber: the natural time capsule.* The Natural History Museum, London. 73 pp.

Schawaller, W. 1978. Neue Pseudoskorpione aus dem Baltischen Bernstein der Stuttgarter Bernstein-sammlung (Arachnida: Pseudoscorpionidea). *Stuttgarter Beiträge zur Naturkunde,* Series B **42**, 1–22.

Schlüter, T. 1990. Baltic Amber. 294–297. *In* Briggs, D. E. G. and Crowther, P. R. (eds.). *Palaeobiology: a synthesis.* Blackwell Scientific Publications, Oxford, xiii + 583 pp.

Schubert, K. 1961. Neue Untersuchungen uuber Bau und Leiben der Bernsteinkiefern [*Pinus succunifera* (Conw.) emend.]. *Beihefte zum Geologischen Jahrbuch* **45**, 1–149.

Selden, P. A. 2002. First British Mesozoic spider, from Cretaceous amber of the Isle of Wight, southern England. *Palaeontology* **45**, 973–983.

Ulmer, G. 1912. Die Trichopteren des baltischen Bernsteins. *Beiträge zur Naturkund Preussens, Königsberg* **10**, 1–380.

Weitschat, W. 1980. *Leben im Bernstein.* Geologisch-Paläontologisches Institut der Universität Hamburg, Hamburg, 48 pp.

Weitschat, W. and Wichard, W. 2002. *Atlas of plants and animals in Baltic amber.* Verlag Dr Friedrich Pfeil, Munich, 256 pp.

Wheeler, W. M. 1915. The ants of the Baltic amber. *Schriften der (Königlichen) Physikalischen-Ökonomischen Gesellschaft zu Königsberg* **55**, 1–11.

Wunderlich, J. 1982. Sex im Bernstein: ein fossiles Spinnenpaar. *Neue Entomologische Nachträge* **2**, 9–11.

Wunderlich, J. 1986. *Spinnen gestern und heute. Fossil Spinnen in Bernstein und ihre heute lebenden Verwandten.* Erich Bauer Verlag, Wiesbaden, 283 pp.

Wunderlich, J. 1988. Die fossilen Spinnen (Araneae) im Baltischen Bernstein. *Beiträge zur Araneologie* **3**, 1–280.

Zompro, O. 2001. The Phasmatodea and *Raptophasma* n. gen., Orthoptera incertae sedis, in Baltic amber (Insecta: Orthoptera). *Mitteilungen der Geologisch-Paläontologisches Institut der Universität Hamburg* **85**, 229–61.

Zompro, O. Adis, J. and Weitschat, W. 2002. A review of the order Mantophasmatodea (Insecta). *Zoologischer Anzeiger* **241**, 269–79.

RANCHO LA BREA

Background: the Pleistocene in North America

By the onset of the Quaternary Period the continents were close to their present positions, although the mid-Atlantic ridge was continuously spreading. At this time, 2.5 million years ago (Bowen, 1999), there was a dramatic deterioration in the climate which over much of North America and northern Europe and Asia remained cold to glacial during the whole of the Quaternary, with temperate to warm intervals of short duration. This is the period of Earth's history known as the 'Great Ice Age', when the climate was the dominant geological force. Icebergs began to appear in northern oceans and vast continental ice-sheets covered much of the northern continents. The North American ice cap covered 13 million square km (5 million square miles) of the continent; it carved the landscape of northern Canada, with meltwaters carrying the debris south as far as the Great Lakes.

The Quaternary Period, lasting for only 2.5 Ma, is much shorter than any other geological period, and is too short to be subdivided on the traditional basis of faunal and floral evolutionary changes. Instead it is subdivided on the basis of climatic changes, which were dramatic at this time. The term 'Ice Age' often gives the wrong impression; the Quaternary was not one continuous glaciation, but was a period of oscillating climate with advances of ice and growth of glaciers punctuated by times when the climate was not very different from that of today.

In North America there were four major cold periods during the Quaternary (compared to six in the British Isles). The Nebraskan, Kansan, Illinoian and Wisconsinan glaciations alternated with intervening warmer periods of the pre-Nebraskan, Aftonian, Yarmouthian and Sangamonian interglacials. The Nebraskan glaciation began around 1 million years ago and lasted about 100,000 years, but it was the final Wisconsinan glaciation, which began about 100,000 years ago, that included the coldest time during the whole of the Ice Age.

Sea level was much lower during the glaciations because millions of cubic kilometres of water from the oceans were turned into ice, lowering sea level eustatically by as much as 120 m (400 ft). The Bering Strait, which today is a shallow sea separating Alaska and Siberia, emerged periodically as a land bridge connecting north-eastern Asia and north-western North America. Although this prevented-exchange of marine organisms between the Pacific and the Arctic oceans, it permitted terrestrial species to migrate between North America and Eurasia. North America species, such as the camel and horse, migrated to Eurasia, while Eurasian mammals, such as mammoths, bison and Man, entered North America. Gradual changes in the North American mammalian fauna resulting from such migration are used to define a succession of North American Land Mammal Ages.

However, at the close of the Wisconsinan glaciation, some time between 12,000 and 10,000 years ago, the mammal faunas across the globe underwent severe changes. In North America 73% of the large mammals (33 genera) became extinct, including mammoths, mastodons, horses, tapirs, camels and ground sloths, together with their predators such as sabre-toothed cats. Whether this was caused simply by climatic changes at the end of the Ice Age, or whether it was the effect of excessive hunting by Man, is still the focus of much debate.

Within the City of Los Angeles, at a locality known as Rancho La Brea (**249**, **250**), can be found one of the world's richest deposits of Ice Age fossils. Preserved in asphalt-rich sediments, so numerous are the remains that they can truly be considered as a Concentration Lagerstätte. Their diversity provides a virtually complete record of life in the Los Angeles Basin between 10,000 and 40,000 years ago, during this vital period at the close of the Ice Age in North America. This exceptional biota defines the Rancholabrean Land Mammal Age (Savage, 1951), and includes approximately 60 different mammal species ranging in size from huge mammoths to the Californian pocket mouse. This virtually entirely preserved ecosystem also includes reptiles such as snakes and turtles, amphibians such as frogs and toads, birds, fish, molluscs, insects, spiders, and numerous plants including microscopic pollen and seeds.

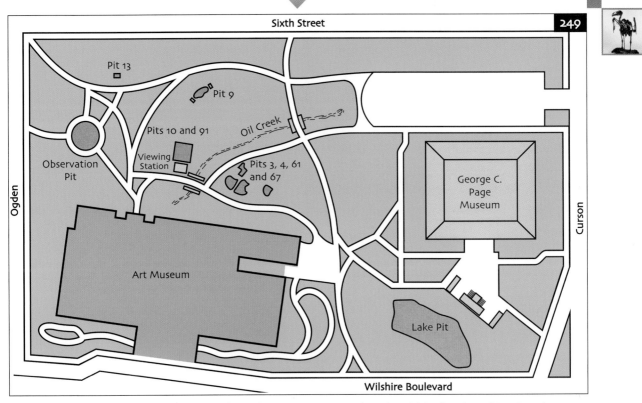

249 Locality map showing the position of Rancho La Brea (Hancock Park) within the City of Los Angeles (after Stock and Harris, 1992).

250 Rancho La Brea, Hancock Park, Los Angeles, showing methane gas bubbling through the oily water of a flooded asphalt working, surrounded by life-size models of Ice-Age mammoths.

251 Naturally occurring asphalt pool at Hancock Park, Los Angeles.

Los Angeles exclusive rights to excavation (Harris and Jefferson, 1985). More than 750,000 bones were removed in the first two years and in 1915 Captain Hancock donated the fossils to the Los Angeles County Museum. At the same time the ranch, later renamed Hancock Park, was also donated to the museum for preservation, research and exhibition. In 1963 Hancock Park was declared a National Natural Landmark (**250**) and in 1969 excavation resumed at Pit 91 in order to recover some of the smaller elements of the fauna and flora, such as insects, molluscs, seeds and pollen, which had hitherto been ignored. Excavation continues to the present day.

The name most associated with the research of these deposits is that of Chester Stock (1892–1950), who was a student of John C. Merriam and who joined some of the early excavations at La Brea from 1913 onwards. He was associated with the Los Angeles County Museum from 1918 until his death and published the first comprehensive monograph of the La Brea fossils (Stock, 1930), which had reached its seventh edition by 1992 (Stock and Harris, 1992).

History of discovery and exploitation of Rancho La Brea

The naturally occurring asphalt in this area (**251**) has been used by man since prehistoric times. The local Chumash and Gabrielino Indians used the sticky 'tar' both as a glue for making weapons, vessels and jewellery, and as waterproofing for canoes and roofing (Harris and Jefferson, 1985), but the first record of these deposits was that of the Spanish explorer Gaspar de Portolá, who noted 'muchos pantamos de brea' (extensive bogs of tar) in 1769. In 1792 José Longinos Martínez recorded '…twenty springs of liquid petroleum' and '…a large lake of pitch…in which bubbles or blisters are constantly forming and exploding' (Stock and Harris, 1992).

In 1828 the area became part of a Mexican land grant known as Rancho La Brea (which literally means 'the tar ranch', although the term 'tar' is not strictly accurate – the naturally occurring bituminous substance derived from petroleum is asphalt). During the nineteenth century, and especially during the 1860s and 1870s, the asphalt began to be mined commercially for road construction and during these operations workers began to find bones, but disregarded these as the remains of recent animals which had become trapped in the sticky bogs.

It was not until 1875, when the owner of the ranch, Major Henry Hancock, presented the tooth of a sabre-toothed cat to William Denton of the Boston Society of Natural History that the true age of the fossils was appreciated. Denton visited the area and collected further specimens of horses and birds. No further interest was shown, however, until 1901, when the Los Angeles geologist W. W. Orcutt visited the area with a view to oil production. His scientific excavations between 1901 and 1905 produced specimens of sabre-toothed cat, wolf and ground sloth which were passed to Dr John C. Merriam of the University of California. Merriam realized the importance of the deposit and excavated between 1906 and 1913, but in that year Captain G. Allan Hancock, the son of Henry Hancock, gave the County of

Stratigraphic setting and taphonomy of the Rancho La Brea biota

Most of the fossils excavated at Rancho La Brea have been estimated by carbon-14 dating to be between 11,000 and 38,000 years old, which means that the sediments in which they are buried were deposited during the final stages of the Wisconsinan glaciation, at the very end of the Pleistocene Epoch (**252**). In terms of the North American Land Mammal Ages, the fauna belongs to the latter part of the Rancholabrean Land Mammal Age, which began 500,000 years ago, defined by the first occurrence of bison in North America.

Prior to the start of the Wisconsinan glaciation, 100,000 years ago, this part of California was submerged by an extended Pacific Ocean. The fall in sea level at the onset of this glaciation exposed a flat plain between the reduced Pacific and the Santa Monica Mountains, on which were numerous interconnected freshwater lakes. Erosion of the mountains by rivers led to the accumulation of fluvial sands, clays and gravels between 12 m (40 ft) and 58 m (190 ft) thick, which gradually raised the level of the plain.

Beneath this alluvial plain Tertiary marine sediments of the Fernando Group, consisting of shales and sandstones interbedded with oil sands, acted as a reservoir for the Salt Lake oilfield. These sediments had been faulted, folded and eroded during the early Pleistocene to form a north-east–south-west trending anticline, and from around 40,000 years ago crude oil began to seep upwards towards the crest of the anticline and into the overlying horizontally bedded Pleistocene fluvial deposits. The lighter petroleum evaporated, leaving sticky pools of natural asphalt at the surface. Many of the asphalt pools within Hancock Park (**251**) are aligned along a north-west–south-east axis, suggesting that the oil seepages may have originated from a subsurface fault.

These shallow asphalt pools formed natural traps for animals and plants especially during the warm summers when the asphalt would have been viscous. Cooler winters may have solidified the asphalt and covered it

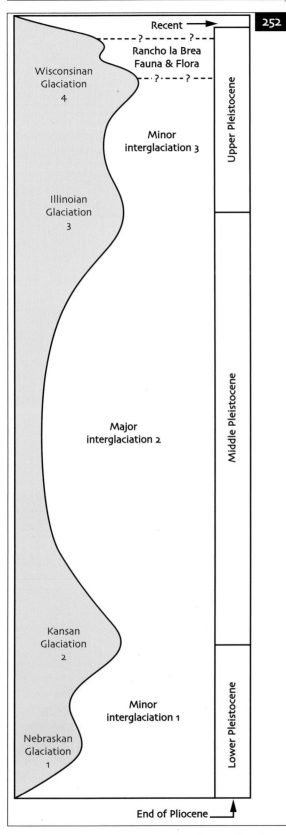

252 Diagram showing the glacial and interglacial stages of the Pleistocene, with the position of the Rancho La Brea fauna indicated (after Stock, 1930).

with river sediments before the trap was reset the following summer. Repetition of this annual cycle produced conical bodies of asphalt (Shaw and Quinn, 1986). Carcasses accumulated in large numbers and it is likely that many scavengers were lured to the pools by an initial victim.

The excellent preservation of the biota seems to be the result not only of rapid burial, but also, and unusually, of impregnation of the bones by asphalt. Soft tissues are generally not present so the deposit cannot be considered as a Conservation Lagerstätte, but the sheer numbers of bones preserved (**253**) define it as a Concentration Lagerstätte; more than 50 wolf skulls and 30 sabre-toothed cat skulls have been collected within just 4 cubic metres (140 cubic ft).

Bones and teeth are preserved almost in their original state, apart from their penetration by oil, which gives a brown or black coloration. Up to 80% of the original collagen is retained (Ho, 1965) and microstructure is well preserved (Doberenz and Wyckoff, 1967). Surface markings on bones still show the positions of nerves and blood vessels and the attachment points of tendons and ligaments. Often oil has accumulated in skull cavities preserving, for example, the tiny bones of the middle ear or even the remains of small mammals, birds and insects. Curiously, epidermal structures are rarely preserved; occasional hairs and feathers are known, but nails and claws of mammals or talons and beaks of birds are not. Chitinous bodies of insects retaining the iridescent colours of wing cases, and fleshy leaves and pine cones, thoroughly impregnated with oil, are not uncommon.

253 Concentration of bones in tar deposits.

Description of the Rancho La Brea biota

Human remains. The skull and partial skeleton of a human female have been recovered and carbon-14 dated at 9,000 years BP, thus postdating the bulk of the biota. 'La Brea Woman' was between 20 and 25 years of age (Kennedy, 1989) and stood approximately 1.5 m (c. 5 ft)) tall. The fractured skull suggests that she may have been murdered and her body dumped in a shallow tar pool (Bromage and Shermis, 1981), although an alternative hypothesis suggests a ritual burial (Reynolds, 1985). Many human artefacts have also been found, mostly less than 10,000 years BP, and include shell jewellery, bone artefacts, wooden hairpins and spear tips.

Dire wolf and other dogs. The dire wolf (*Canis dirus*) is the most common mammal from La Brea, known from over 1,600 individuals (**254**, **255**). They probably fed in packs on animals stuck in the tar and became trapped themselves. The large head with strong jaws and massive teeth made it the major predator of La Brea. Other dogs include the grey wolf (*Canis lupus*) and coyote (*Canis latrans*), which is the third most common La Brea mammal, and which is slightly larger than the modern representatives of this species (**256**). Domestic dogs are also present, one being associated with the skeleton of La Brea Woman (Reynolds, 1985).

Sabre-toothed cat and other cats. The state fossil of California, *Smilodon fatalis*, is the best known of all the La Brea mammals and is the second most common (**257**, **258**). About the same size as the African lion, the exact function of the large upper canine teeth is disputed. Conventionally believed to have been used to stab and kill their prey, recent studies suggest that their fragility would have been more suited to slicing open the soft underbelly of the prey after the kill (Akersten, 1985). Other large cats include the American lion, puma, bobcat and jaguar.

Elephants. The imperial mammoth, *Mammuthus imperator*, was the largest of the La Brea mammals, standing almost 4 m (13 ft) tall and weighing almost 5,000 kg (5 tons) (**259**, **260**). The American mastodon, *Mammut americanum*, was smaller, at 1.8 m (6 ft) tall (**261**) and fed on leaves and twigs, unlike the mammoth, which fed on grass.

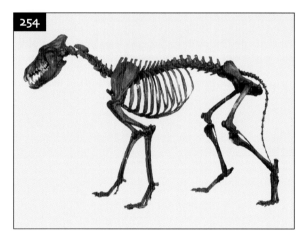

254 Skeleton of the dire wolf, *Canis dirus* (GCPM). Length of body 1–1.4 m (3–4.5 ft).

255 Reconstruction of dire wolf.

256 Reconstruction of coyote.

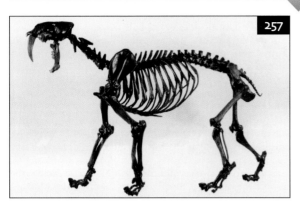

257 Skeleton of the sabre-tooth cat *Smilodon fatalis* (GCPM). Length of body 1.4–2 m (4.5–6.5 ft).

258 Reconstruction of sabre-toothed cat.

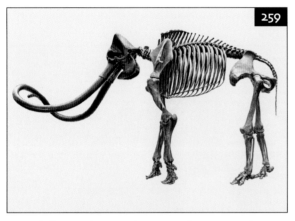

259 Skeleton of imperial mammoth, *Mammuthus imperator* (GCPM). Height up to 4 m (13 ft).

260 Reconstruction of mammoth.

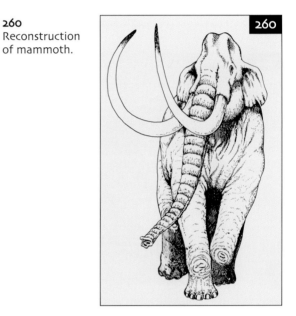

261 Reconstruction of mastodon.

Ground sloth. This large, primitive mammal, *Glossotherium harlani*, is related to the modern South American tree sloth, but its flat grinding teeth suggest that it fed on grass. At 1.8 m (6 ft) tall these animals had dermal ossicles embedded into the skin of their necks and backs as a protection against predators, similar to the ankylosaur dinosaurs (**262**, **263**).

Other large mammals. Additional carnivores include three species of bear – the short-faced bear, black bear and grizzly bear. Herbivores are diverse and include bison (**264**), horses, tapirs, peccaries, camels, llamas, deer and pronghorns (**265**).

Small mammals. These include the carnivorous skunks, weasels, racoons and badgers; the insectivorous shrews and moles; numerous rodents including mice, rats, gophers and ground squirrels; the lagomorphs (jackrabbits and hares); and bats.

Birds. The protective asphalt coating has preserved more fossil birds than at any other location in the world. Many were predators or scavengers, such as condors, vultures and teratorns, which became trapped while feeding on carcasses. The extinct, raptor-like teratorn, *Teratornis merriami*, was 0.75 m (2.5 ft) tall with a wing-span of 3.5 m (11.5 ft), one of the largest known flying birds (**266**). The large number of water birds, including herons, grebes, ducks, geese and plovers, may have landed on the asphalt, mistaking its reflective surface for a pond. There are more than 20 species of eagles, hawks and falcons, with the Golden Eagle being the most common bird. Storks, turkeys, owls and numerous smaller songbirds complete the bird fauna.

Reptiles, amphibians and fish. Seven different lizards (Brattstrom, 1953), nine snakes (La Duke, 1991a), one pond turtle, five amphibians (including toads, frogs,

262 Skeleton of ground sloth *Glossotherium harlani* (GCPM). Height 1.8 m (6 ft).

263 Reconstruction of ground sloth.

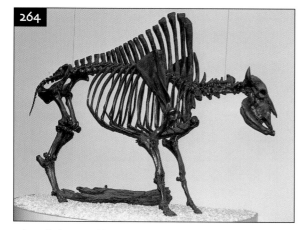

264 Skeleton of bison *Bison antiquus* (GCPM). Height 2.1 m (7 ft).

265 Reconstruction of pronghorn.

tree frogs and climbing salamanders) (La Duke, 1991b), and three species of fish (rainbow trout, chub and stickle-back) (Swift, 1979) are known from La Brea, which suggests that permanent bodies of water existed in this area.

Invertebrates. Freshwater molluscs (five bivalves and fifteen gastropods) suggest the presence of ponds or streams at least during part of the year. In addition eleven terrestrial gastropod species, which would have lived on leaf litter, have been recognized. Seven orders of insect include grasshoppers and crickets, termites, true bugs, leafhoppers, beetles, flies, ants and wasps. Scorpions, millipedes and several spiders are also known. Many of these were terrestrial and are rarely found fossilized. Some of the beetles and flies would have become trapped while feeding on carrion. Others were blown onto the sticky asphalt or became stuck when crawling over it.

Plants. Fossil plants from La Brea include wood, leaves, pine cones, seeds and microscopic pollen and diatoms.

The fauna and flora from Rancho La Brea is well documented and illustrated by Harris and Jefferson (1985) and Stock and Harris (1992).

Palaeoecology of the Rancho La Brea biota

The fauna and flora of the Rancho La Brea 'tar pits' represent a terrestrial ecosystem situated on the western coastal plains of the North American continent, between 30° and 35°N, as at present, but with a cooler, glacial climate. The flora includes many plants that no longer occur in this region and indicates that while glacial winters may have resembled modern winters, the summers were not only cooler, but were also more moist (Johnson, 1977a, b). Annual rainfall was probably twice

that of today; freshwater ponds and streams covered the plains, supporting fish, turtles, frogs, toads, molluscs and aquatic insects, and a rich vegetation included four distinct assemblages.

The slopes of the Santa Monica Mountains were covered with chaparral – tall, densely packed woody bushes of lilac, scrub oak, walnut and elderberry – while the deep, protected canyons were home to redwood, dogwood and bay. The rivers were lined with sycamore, willow, alder, raspberry, box elder and live oak, but the plains were covered with coastal sage scrub (drought-tolerant woody bushes) interspersed with wide areas of grass and herbs, with occasional groves of closed-cone pine, valley oak, juniper and cypress (Harris and Jefferson, 1985).

These wide plains supported large herds of hoofed mammals feeding on the rich vegetation – bison, horse, ground sloths, camels, pronghorns and mammoths, with occasional visits by peccaries, deer, tapirs and mastodons. These numerous herbivores in turn supported a diverse population of cursorial carnivores, including the various species of dogs, cats and bears. There is no need to infer a radically different climate from that of the present to account for the variety of Pleistocene mammals in this region. In fact the similarity of the smaller mammals to the present day fauna (rodents, rabbits, shrews), plus the absence of mammals that are elsewhere associated with very cold climates (e.g. musk oxen), suggest that conditions were not significantly colder during the Pleistocene.

In all more than 600 species have been recorded from La Brea, comprising approximately 160 plants and 440 animals. These include 59 species of mammals (represented by over one million fossils) and 135 different birds (represented by 100,000 specimens). The most significant statistic, however, is the disproportionate percentage of carnivores over herbivores, which does not conform to the normal pyramidal structure of a biological ecosystem in which carnivores form the top of the pyramid, outnumbered by herbivores.

The Rancho La Brea assemblage clearly does not represent an accurate cross section of the ecosystem. The most obvious explanation is that a single herbivore, on becoming trapped in the asphalt, might be pursued by a whole pack of carnivores. When these also became trapped, both would attract additional scavengers to their carcasses. The dogs, cats and bears, plus the smaller carnivores, such as skunks, weasels and badgers, constitute 90% of the mammal fauna, while the herbivores make up only 10%. Similarly the carnivorous birds (condors, vultures, teratorns, eagles, hawks, falcons and owls) comprise approximately 70% of the bird fauna.

A further bias in the La Brea fauna is the preponderance for obvious reasons of young, aged and maimed individuals. There may also be a sampling bias as early excavators concentrated on collecting the larger, more impressive, mammals. Given that the preservation potential of a mammal becoming stuck in the asphalt is very high, it is a sobering fact that an entrapment episode of a single herbivore followed by, say, four dire wolves, a sabre-toothed cat and a coyote need only occur once every decade, over a period of 30,000 years, to account for the number of mammals represented in the collections (Stock and Harris, 1992).

266 Skeleton of the teratorn *Teratornis merriami* (GCPM). Height 750 mm (2.5 ft); wing-span 3.5 m (10.5 ft).

Comparison of Rancho La Brea with other Pleistocene sites
Permafrost of Siberia and Alaska

Many animals and plants survived the glaciations by occupying slightly warmer areas to the south of the ice sheets on both the North American and Asian continents. The Russian mammoth steppe was a vast grassy tundra fringing the northern ice sheet; its climate was too dry to support a large build-up of ice, and it was effectively a frozen version of today's hot African grasslands. It was home, 40,000 years ago, to woolly mammoths, woolly rhinos, bison, giant Irish deer, horses, and large predatory cats and although there was a summer thaw, allowing some vegetation, the subsurface was permanently frozen. This gave rise to what must be considered the ultimate Fossil-Lagerstätte – the deep-freeze of the Siberian permafrost.

Woolly mammoths and rhinos in particular were sometimes engulfed in bogs and deep-frozen in the permafrost, where they have remained ever since. Desiccation by freezing, where the moisture is not released to the atmosphere, but forms ice crystals around the mummy, causes the carcass to shrink and shrivel as it dries. (This should not be confused with freeze-drying, where moisture is removed by sublimation and the carcass retains its original form; see Guthrie, 1990.) Carcasses are thus mummified (compare with Baltic amber, Chapter 13), and not only is the coarse outer hair and soft downy inner hair perfectly preserved, the meat is also so fresh that it has been eaten by dogs, and apparently also by humans. One of the first such mammoths to be excavated and examined scientifically was that from Beresovka in Siberia in 1900. According to Kurtén (1986) the excavators attempted to eat the 40,000 year-old meat, but were 'unable to keep it down, in spite of a generous use of spices'. This mammoth is preserved with its final mouthful of food still in its mouth, and is on display at the Zoological Museum in St Petersburg.

Several mammoths, woolly rhinos, bison, horse and musk ox have been recovered from the permafrost since the 1970s, one of the most well known being the complete baby woolly mammoth (*Mammuthus primigenius*) found at Magadan in Siberia in 1977. The young male (christened 'Dima') was found beneath 2 m (6 ft) of frozen silt and is approximately 40,000 years old. Similar remains have been found in the permafrost of Alaska. In 1976 a partial baby mammoth, with a rabbit, a lynx and a lemming (or vole), was discovered in the Fairbanks district (Zimmerman and Tedford, 1976) and dated by carbon-14 at 21,300 years B.P. The skin, hair and eyes of the mammoth were well preserved and the liver of the rabbit was recognizable after rehydration, but most of the internal organs were decayed and replaced by bacteria.

Unfortunately, hopes of bringing the mammoths back to life using modern cloning techniques, by inserting mammoth DNA into the empty egg cell of an elephant, are some way off. As with insects preserved in amber (Chapter 13), dehydration over thousands of years has destroyed the chains of DNA and only small parts of the chains remain.

Further Reading

Akersten, W. A. 1985. Of dragons and sabertooths. *Terra* **23**, 13–19.

Bowen, D. Q. 1999. A revised correlation of Quaternary deposits in the British Isles. *Geological Society Special Report* **23**, 1–174.

Brattstrom, B. H. 1953. The amphibians and reptiles from Rancho La Brea. *Transactions of the San Diego Society of Natural History* **11**, 365–392.

Bromage, T. G. and Shermis, S. 1981. The La Brea Woman (HC 1323): descriptive analysis. *Society of California Archaeologists Occasional Papers* **3**, 59–75.

Doberenz, A. R. and Wyckoff, R. W. G. 1967. Fine structure in fossil collagen. *Proceedings of the National Academy of Sciences* **57**, 539–541.

Guthrie, R. D. 1990. *Frozen fauna of the mammoth steppe.* University of Chicago Press, Chicago.

Harris, J. M. and Jefferson, G. T. 1985. Rancho La Brea: treasures of the tar pits. *Natural History Museum of Los Angeles County, Science Series* **31**, 1–87.

Ho, T. Y. 1965. The amino acid composition of bone and tooth proteins in late Pleistocene mammals. *Proceedings of the National Academy of Sciences* **54**, 26–31.

Johnson, D. L. 1977a. The Californian Ice-Age refugium and the Rancholabrean extinction problem. *Quaternary Research* **8**, 149–153.

Johnson, D. L. 1977b. The late Quaternary climate of coastal California: evidence for an Ice Age refugium. *Quaternary Research* **8**, 154–179.

Kennedy, G. E. 1989. A note on the ontogenetic age of the Rancho La Brea hominid, Los Angeles, California. *Bulletin of the Southern California Academy of Sciences* **88**, 123–126.

Kurtén, B. 1986. *How to deep-freeze a mammoth.* Columbia University Press, New York, vii + 121 pp.

La Duke, T. C. 1991a. The fossil snakes of Pit 91, Rancho La Brea, California. *Contributions in Science* **424**, 1–28.

La Duke, T. C. 1991b. First record of salamander remains from Rancho La Brea. *Abstract of the Annual Meeting of the California Academy of Science* **7**.

Reynolds, R. L. 1985. Domestic dog associated with human remains at Rancho La Brea. *Bulletin of the Southern California Academy of Sciences* **84**, 76–85.

Savage, D. E. 1951. Late Cenozoic vertebrates of the San Francisco Bay region. *University of California Publications in Geological Sciences* **28**, 215–314.

Shaw, C. A. and Quinn, J. P. 1986. Rancho La Brea: a look at coastal southern California's past. *California Geology* **39**, 123–133.

Stock, C. 1930. Rancho La Brea: a record of Pleistocene life in California. *Natural History Museum of Los Angeles County, Science Series* **1**, 1–84.

Stock, C. and Harris, J. M. 1992. Rancho La Brea: a record of Pleistocene life in California. *Natural History Museum of Los Angeles County, Science Series* **37**, 1–113.

Swift, C. C. 1979. Freshwater fish of the Rancho La Brea deposit. *Abstracts, Annual Meeting Southern California Academy of Science* **88**, 44.

Zimmerman, M. R. and Tedford, R. H. 1976. Histologic structures preserved for 21,300 years. *Science* **194**, 183–184.

MUSEUMS AND SITE VISITS

Chapter 1 Ediacara
Museums
1. South Australian Museum, Adelaide, Australia.
2. Western Australian Museum, Perth, Australia.
3. Sedgwick Museum, University of Cambridge, Cambridge, England.
4. University of California Museum of Paleontology, Berkeley, California, USA (online exhibit: http://www.ucmp.berkeley.edu/).
5. Humboldt State University Natural History Museum, Arcata, California, USA.
6. Manchester University Museum, Oxford Road, Manchester, England.

Sites
Ediacaran biota localities are situated in the Flinders Ranges some 400 km (250 miles) north of Adelaide. There is a bitumen road running along the western side of the Flinders as far as Lyndhurst, and Wilpena can be reached by a made road; all other routes in the area are gravel. The area around Wilpena is in a National Park, where there is tourist accommodation and 4WD tours which include some of the geological sites.

Chapter 2 The Burgess Shale
Museums
1. National Museum of Natural History, Smithsonian Institution, Washington DC, USA.
2. Royal Ontario Museum, Toronto, Canada.
3. Field Visitor Center, Field, British Columbia, Canada.
4. Royal Tyrell Museum, Drumheller, Alberta, Canada.
5. Sedgwick Museum, University of Cambridge, Cambridge, England.
6. Manchester University Museum, Oxford Road, Manchester, England.

Sites
Walcott's Burgess Shale Quarry is situated in Yoho National Park and visits to it are strictly controlled. Guided trips may be booked in advance by calling the Yoho–Burgess Shale Research Foundation (250–343–6480). The hike is strenuous, involving some steep climbs, and should only be undertaken if you are fit and healthy. However, collection of fossils is absolutely forbidden and there are severe penalties for removing fossils from the Park. The Burgess Pass–Yoho Pass hiking trail passes close to the quarries, and in good weather provides a superb day in spectacular scenery. Note that the quarry is visible through good binoculars from Emerald Lake Lodge, situated on the moraine damming Emerald Lake. For further information apply to the Superintendent, Yoho National Park, PO Box 99, Field, British Columbia, Canada. Telephone 250–343–6324.

Chapter 3 The Soom Shale
Museums
The sparse Soom Shale fauna is deposited in the Geological Survey of South Africa and is not on exhibition anywhere at present.

Sites
The Soom Shale quarry at Keurbos Farm is situated some 13 km (8 miles) south of Clanwilliam on the dirt road heading towards Algeria. Care should be taken on this road, especially in wet weather when it is rather slippery, because 4WD farm traffic uses it at greater speeds than ordinary cars. The quarry (**42**) is unmistakable, being on the east side of the road where a bend crosses over a small valley about 2 km (1.25 miles) north of Keurbos. There is a detailed locality map in Theron *et al.* (1990). There is little to see apart from the grey shale itself. However, the surrounding area has spectacular sandstone scenery (**40**), and Clanwilliam is a wonderful centre for exploring the northern Cedarberg.

Chapter 4 The Hunsrück Slate
Museums

1. Lehr-und Forschungsgebiet für Geologie und Paläontologie der Rheinisch Westfälischen Technischen Hochschule, Aachen, Germany.
2. Schloßparkmuseum und Römerhalle, Bad Kreuznach, Germany.
3. Hunsrückmuseum, Simmern, Germany.
4. Museum für Naturkunde der Humboldt-Universität zu Berlin, Germany.
5. Institut für Paläontologie der Universität Bonn, Bonn, Germany.
6. Hunsrück-Fossilienmuseum, Bundenbach, Germany.
7. Museum für Naturkunde der Stadt, Dortmund, Germany.
8. Naturhistorisches Museum Mainz, Mainz, Germany.
9. Naturmuseum und Forschungsinstitut Senckenberg, Frankfurt am Main, Germany.
10. Bergbaumuseum, Bochum, Germany (Bartels Collection).

Sites

The most spectacular pyritized fossils come from the area around the small villages of Bundenbach and Gemünden in the Hahnenbach and Simmerbach valleys to the southwest of Koblenz. The Hunsrück-Fossilienmuseum in Bundenbach has an adjacent disused slate mine open for visitors from April through September. Outside the mine the extensive spoil tips give excellent access for fossil hunting. Fossils are not uncommon, but are difficult to recognize in the field and require expert preparation. Another good outcrop is the Eschenbach-Bocksberg Quarry to the southwest of Bundenbach. It is necessary to contact the owner, Johann Backes (info@Johann-Backes.de) to make an appointment. Excursions for individuals and groups are also arranged by local geologist Wouter Südkamp (Gartenstraße 11, D-55626 Bundenbach; fax + 49 6544 9093; website www.hunsrueck.com/suedkamp). From Gemünden a 4 km (2.5 miles) nature trail, Geologischer Hunsrück-Lehrpfad, can be followed from the village, and includes a viewing platform above the extensive spoil heaps of the opencast Kaiser Mine. The mine is open to visitors from April through September (telephone + 49 6765 1220), but is closed on Mondays, as are most of the museums in the area.

Chapter 5 The Rhynie Chert
Museums

Being generally microscopic, Rhynie plants and animals make poor fossils for display purposes, but there are some good online exhibitions at the University of Aberdeen Geological Collections (http://www.abdn.ac.uk/rhynie/) and the University of Münster Palaeobotanical Research Group (http://www.uni-muenster.de/GeoPalaeontologie/ Palaeo/ Palbot/erhynie.html). The National Museum of Scotland, Edinburgh, Scotland, has a display on the Rhynie Chert.

Sites

There is no outcrop of the chert, only loose blocks in a field and incorporated into stone walls and, indeed, most of these have now been collected. Material from a trench dug in the 1970s is stored in crates in the Natural History Museum, London. A pilgrimage to the little village of Rhynie and a walk to the field (**72**) are worthwhile because this is a pleasant and little-visited part of Scotland.

Chapter 6 Mazon Creek
Museums

1. Field Museum of Natural History, Chicago, Illinois, USA.
2. National Museum of Natural History, Smithsonian Institution, Washington DC, USA.
3. Illinois State Museum, Springfield, Illinois, USA (online exhibit: http://www.museum.state.il.us/exhibits/ mazon_creek/).
4. Burpee Museum of Natural History, Rockford, Illinois, USA.

Sites

Collection at the Mazon Creek sites is best organized through the Mazon Creek Project: a group of amateur and professional palaeontologists interested in education and scientific research on Mazon Creek fossils. The project holds open houses, leads field trips, and provides information on the Mazon Creek fossils. Contact: The Mazon Creek Project, Northeastern Illinois University, Department of Earth Sciences, 5500 N. St Louis Ave., Chicago, IL 60625 (Telephone 773-442-5759).

Chapter 7 Grès à Voltzia
Museums

The Grès à Voltzia fossils are held in the Université Louis Pasteur de Strasbourg and are not on display.

Sites

The Grès à Voltzia sandstone is quarried extensively in the northern Vosges Mountains (**119**). Permission from the quarry owner should be sought before entering any of these quarries, working or not.

Chapter 8 The Holzmaden Shale
Museums

1. Urwelt-Museum Hauff, Holzmaden, Germany.
2. Urweltsteinbruch Museum, Holzmaden, Germany.
3. Staatliches Museum für Naturkunde, Stuttgart, Germany.

Sites

Some of the quarries around Holzmaden and Ohmden are open to collectors on payment of a small fee. These villages are approximately 40 km (25 miles) south-east of Stuttgart and are accessed via the A8 Stuttgart–Munich autobahn. Leave the autobahn at the Aichelberg exit and follow the clear signs (with the crocodile *Steneosaurus*!) to the Urwelt-Museum Hauff in Holzmaden. Immediately opposite this museum is the Urweltsteinbruch Museum, which has a small quarry attached. Hammers and chisels are provided, but the beds here are not particularly fossiliferous. More productive is Schieferbruch Kromer at Ohmden, which is open on Mondays to Saturdays from April to October (telephone/fax + 49 7023 4703). The Urwelt-Museum Hauff will arrange visits to the latter.

Chapter 9 The Morrison Formation
Museums

1. American Museum of Natural History, New York, USA.
2. Museum of the Rockies, Montana State University, Bozeman, Montana, USA.
3. Field Museum of Natural History, Chicago, Illinois, USA.
4. Carnegie Museum of Natural History, Pittsburgh, Pennsylvania, USA.
5. Geological Museum, University of Wyoming, Laramie, Wyoming, USA.
6. Black Hills Institute of Geological Research, Hill City, South Dakota, USA.
7. Dinosaur National Monument, Vernal, Utah, USA.
8. The Wyoming Dinosaur Center, Thermopolis, Wyoming, USA.
9. Fossil Cabin Museum, Como Bluff, Medicine Bow, Wyoming, USA.
10. The Natural History Museum, London, England.
11. Saurier Museum, Aathal, Switzerland.
12. Museum für Naturkunde der Humboldt-Universität zu Berlin, Germany.

Sites

There are many places to view the Morrison Formation, covering as it does such a vast area of the western United States. Most spectacular is Dinosaur National Monument in Utah, while for organized collecting the Wyoming Dinosaur Center and Dig Sites is excellent and accessible. Dinosaur National Monument was originally established to protect a quarry containing 1,600 exposed dinosaur bones from 11 different species of dinosaur. The Quarry Visitor Center is 11 km (7 miles) north of Highway 40, about 32 km (20 miles) from Vernal, Utah. The Wyoming Dinosaur Center and Dig Sites at Thermopolis, in the Big Horn Basin of Wyoming, consists of a new museum exhibiting many specimens collected on the adjacent land (including a complete skeleton of *Camarasaurus*, nicknamed 'Morris'). Interpretive dig site tours visit several active collecting sites on an adjacent 6,000 ha (15,000 acre) ranch, while the 'dig-for-a-day' programme allows visitors to work alongside professional palaeontologists in the field (website www.wyodino.org).

Chapter 10 The Solnhofen Limestone
Museums

1. Bayerische Staatssammlung für Paläontologie und historische Geologie, München, Germany.
2. Bürgermeister Müller Museum, Solnhofen, Germany.
3. Carnegie Museum of Natural History, Pittsburgh, USA.
4. Jura-Museum, Eichstätt, Germany.
5. Museum auf dem Maxberg, Solnhofen, Germany.
6. Museum Berger, Harthof, Germany.
7. Museum für Geologie und Mineralogie, Dresden, Germany.
8. Museum für Naturkunde der Humboldt-Universität zu Berlin, Germany.
9. Naturkunde-Museum Bamberg, Germany.
10. Naturmuseum Senckenberg, Frankfurt, Germany.
11. Staatliches Museum für Naturkunde, Karlsruhe, Germany.
12. Staatliches Museum für Naturkunde, Stuttgart, Germany.
13. Teylers Museum, Haarlem, Netherlands.
14. American Museum of Natural History, New York, USA.
15. Manchester University Museum, Oxford Road, Manchester, England.
16. The Natural History Museum, London, England.

Sites

Many of the working quarries exclude fossil hunters. However some quarries are accessible, such as the one adjacent to the small, private Museum Berger at Harthof. Head north-west from Eichstätt on the B13 across the Altmühl valley towards Weißenburg. After a few kilometres turn left towards Schernfeld and then almost immediately left again to Harthof. The museum displays some excellent Solnhofen fossils and on payment of a small fee you can visit the adjacent working limestone quarry to collect. The floating crinoid *Saccocoma* is common; other fossils are not, but remember that the Berlin *Archaeopteryx* was found here in 1877! Hard hats are not required, but the quarry is often very muddy.

Chapter 11 The Santana and Crato Formations
Museums

1. Museu Nacional, Rio de Janeiro, Brazil.
2. Museu de Paleontologia da Universidade Regional do Cariri, Santana do Cariri, Ceará, Brazil.
3. Departamento Nacional da Produção Mineral (DNPM), Crato, Ceará, Brazil.
4. Small private museum in Jardim, Ceará, Brazil.
5. Museum für Naturkunde der Humboldt-Universität zu Berlin, Germany.
6. Staaliches Museum für Naturkunde, Karlsruhe, Germany.
7. Staaliches Museum für Naturkunde, Stuttgart, Germany.
8. American Museum of Natural History, New York, USA.

Sites

The book by David Martill published by the Palaeontological Association (1993) gives good advice on the practicalities and logistics of visiting field sites in both the Crato and Santana Formations in the State of Ceará in north-east Brazil. There are many flights to Rio de Janeiro from where it is possible to hire a car or endure a 36-hour coach journey to the north-east. Flying to Fortaleza or Recife cuts out about two days of the three-day drive from Rio. There are hotels in Crato or much cheaper Pousada accommodation in the village of Nova Olinda, which is in walking distance of several small quarries in the Crato Formation. Martill (1993) also gives several field itineraries; the Santana Formation and its fossils are best seen around the villages of Cancau, near Santana do Cariri on the north of the plateau, and Jardim, on the south.

It is important to realize that, while it is not illegal to collect fossils in Brazil, it is illegal to buy them – all commercial trade in fossils is forbidden in Brazil. You will, however, be offered fossils very cheaply by small children who work the Crato Plattenkalks by hand, and by 'fish miners', many of whom live in adobe huts in the poor villages. It is also illegal to export fossils from Brazil unless you have the appropriate authorization from the relevant government department. It is possible to acquire this from the Departamento Nacional da Produção Mineral (DNPM) in Crato, but such authorization is very tightly controlled and is only ever given for bona fide research or museum display.

Chapter 12 Grube Messel
Museums

1. Naturmuseum und Forschungsinstitut Senckenberg, Frankfurt am Main, Germany.
2. Fossilien- und Heimatmuseum, Messel, Germany.
3. Hessische Landesmuseum, Darmstadt, Germany.
4. Staatliche Museum für Naturkunde, Karlsruhe, Germany.

Sites

For visits to the Messel quarry, which is a UNESCO World Heritage Site, permission and a guide must be obtained from: Museumsverein Messel e.V., Albert-Schweitzer-Straße 4a, 64409 Messel, Germany (Telephone 06159 / 5119).

Chapter 13 Baltic Amber
Museums

1. The Natural History Museum, London, England.
2. Museum of the Earth, Warsaw, Poland.
3. The Zoological Institute, St Petersbourg, Russia.
4. Museum für Naturkunde der Humboldt-Universität zu Berlin, Germany.
5. Staaliches Museum für Naturkunde, Stuttgart, Germany.
6. Swedish Amber Museum, Höllviken, Sweden.
7. The Amber Museum, Palanga, Lithuania.
8. Amber Museum Gallery, Nida and Vilnius, Lithuania.
9. Ravmuseet, Oksbøl, Denmark.
10. Skagen Ravmuseum, Skagen, Denmark.
11. Geological Museum, Copenhagen, Denmark.
12. Kaliningrad Amber Museum, Kaliningrad, Russia.
13. Yantarnyi Amber Mine Museum, Kaliningrad, Russia.
14. American Museum of Natural History, New York, USA.

Sites

The Yantarny opencast amber mine and museum figures on most tourist excursions to the Kaliningrad region. If you go under your own steam, there are six trains a day from Kaliningrad to Yantarny; the journey takes about an hour. You can buy pieces of the locally produced gold and silver amber jewellery at the factory outlet. Otherwise, amber can be collected from many parts of the Baltic and North Sea coasts, from Russia to East Anglia. Being light, it tends to collect along the tide-line. It is considered best to look after storms.

Chapter 14 Rancho La Brea
Museum

George C. Page Museum of La Brea Discoveries, Hancock Park, Los Angeles, USA.

Sites

The 9 ha (23 acre) Hancock Park is situated between Wilshire Boulevard and 6th Street, 11 km (7 miles) west of the civic center of Los Angeles. Entrance to the park is free and several 'tar pits' can be observed – relics of sites excavated for asphalt and for fossils. A large lake, surrounded by life-sized reproductions of Pleistocene elephants, represents the flooded site of a former commercial working for asphalt, with methane gas constantly bubbling through the oily water. A special Observation Pit at the west end of the park surrounds a partially excavated deposit of fossils and asphalt. Collecting is not possible, but the George C. Page Museum, opened in 1977 at the east end of the park, houses more than two million specimens collected from this site.

INDEX

Note: page numbers in **bold** text refer to figures or tables in the text where these are not included within page ranges